英語で学ぶ計算理論
Theory of Computation

Thomas Zeugmann
湊　　真一　　共著
大久保　好章

コロナ社

まえがき

　本書は，簡潔でなおかつ包括的な**計算理論**の入門書である．対象とする読者は，計算機科学(コンピュータサイエンス)を学ぶすべての学生であり，応用計算機科学，ソフトウェア工学，計算機メディア技術，人工知能，知識発見や，理論を含む計算機科学の諸分野のいずれを専門とするかは問わない．さらに，大学院学生にとっては日常の参考書として，あるいは，英文の研究論文を初めて執筆する際に役立つと思われる．

　本書は15の章からなり，(学部や)大学院での半期の講義に適したものとなっている．冒頭で述べたように，計算理論は，理論と実用とが出会う場である．今日，多くのソフトウェアパッケージは，互いに関連し合うおびただしい数のプログラムからなるが，それらが，なぜ，そして，どのように動作するかの理解を深めることは重要である．

　すべてを**明解**に，**簡潔**に，かつ，**正確**に表現するには，しっかりとした形式化が必要である．本書の執筆に当たっては，離散数学の基礎的な知識を持つ読者を想定している．特に，数学的な主張を証明するための手段に関する基礎的理解があることが望ましい．その他の必要な事柄については，本文中で説明を与えている．さらに，基本的な概念については例示により説明している．形式的な正確さと直感的なアイディアをうまく両立させることによって，学生と指導者がともに，本書に示されている多くの素晴らしい見識を享受できるものと確信している．

　われわれが最も強調したいことは，情報学における理論と実用の相互作用である．理論は実用上重要な問題を扱うべきであり，一方，実用に関する情報学は，最大限の能力を発揮するための確固たる基礎を必要とする．したがって，本書の前半2/3は，形式言語とオートマトン理論にあてている．形式言語は，

様々な場面で出現するため，応用計算機科学にとって必要不可欠である．よって，本書では，文法 (生成の形式化)，オートマトン (受理の形式化)，および正規言語・文脈自由言語に対する文法とオートマトンの関連性を扱っている．

後半の 1/3 では，部分帰納的関数とチューリング機械を通して，アルゴリズムに対する直感的な理解の定式化を行う．これら二つのモデルの等価性を示し，万能チューリング機械の存在，すなわち，**すべての可能な計算を実行できるただ一つの計算装置**が存在することを証明する．その後，いかなる計算機をもってしてもまったく解くことのできない問題が存在することを示す．さらに，停止問題，ポストの対応問題の議論を経て，この理論を適用することによって，このような形式言語理論における自然な問いに対する，より完全な全体像を明らかにする．最後に，不動点定理，再帰定理，およびライスによる有名な定理を示した後，(抽象的な) 計算量クラスについて述べ，それらのいくつかの性質について学ぶ．

さらに本書は，チョムスキー-シュツェンベルガーの定理，部分帰納的関数の解説，デデキントの正当化定理，抽象的計算量クラスとそれらの性質といった，他書ではほとんど扱われない重要な事柄も含んでいる．

なお，本書には講義ですぐに使える 15 回分のスライド (日本語版・英語版) が付属する．この教材は，北海道大学において実際に使われたものである．これらスライドおよび本書は，自ら興味のある章を学ぶことによって自身の英語スキルを高めたい日本人学生や，日本の大学で計算機科学の勉強を修めたい外国人留学生に向けたものである．

われわれのこれまでの経験から，計算理論のような高度に専門化された話題を扱う際には，正確な専門用語を見つけることは，しばしば困難を伴う．こうした作業を容易にすべく，本文中に用語の和訳を付けてある．重要な専門用語が導入される際には，イタリックで**強調され**，直後に適切な日本語訳を挿入している．

定義文 (Definition) は全体がイタリック体で記述され，その中で定義される用語のみローマン体となっている．定義される用語にも日本語訳が付されている．

また本書のような教科書は，もちろん参考書としても利用される．そこで，通常の索引，および，日本語索引の二種類の索引を用意した．

書籍は，生身の教師のように，学生と話をしたり質問に答えたりすることはできない．そのような不便さをカバーするため，課題 (problem) と演習 (exercise) を付けることとした．課題と演習では難易度については大きな差はない．唯一の違いは，課題には解答が付いている (添付 CD-ROM) が，演習には解答が付いていないという点である．これらの課題と演習は，本文に加えてさらに多くの興味深い事柄について光をあてるものであり，セミナーで利用したり，あるいは宿題として課したりするとよい．なお，本書で示した解答は，唯一のものではないことにも注意してもらいたい．この点を強調するために，いくつかの問題に対しては別解を与えている．

最後になるが，本書に沿った学修は，おそらく相当に努力を要するものになると思われる．しかし，それはウィリアム・クラーク博士 (北海道大学建学の父，札幌農学校初代教頭) の激励に沿ったものである．

    Boys, be ambitious !   少年よ，大志を抱け !
もちろん今日では，われわれは次のとおり言い替えよう．
    Girls and Boys, be ambitious ! 少年少女よ，大志を抱け !

2009 年 2 月

                ツォイクマン トーマス
                湊 真一
                大久保 好章

Preface

This book provides a comprehensive albeit short introduction into the *theory of computation*. The intended audience are all students of computer science independently of whether they aim to specialize in applied computer science, software engineering, computer-based media technologies, artificial intelligence, knowledge discovery, or any branch of computer science including theory. Graduate students may find this book also useful as a reference or when writing their first own research papers.

The text is organized in 15 chapters and thus well suited for a one semester (under)graduate course. As said above, theory of computation is the place where theory and practice meet. Though nowadays many software packages include an impressive amount of the relevant programs, it is important to gain an understanding why and how everything works.

Expressing everything *clearly, concisely* and *correctly* requires a certain degree of formalization. We assume basic knowledge of discrete mathematics, only. In particular, it may be helpful if the reader has a basic understanding of what constitutes a proof of a mathematical assertion. All the remaining material is introduced in the text. Fundamental concepts are exemplified, too. We are strongly convinced that a solid compromise between formal correctness and intuitive ideas may help both the students and instructors to enjoy the wealth of insight this book is aiming to present.

Main emphasis is put on the interaction of theory and practice in informatics. Theory must deal with problems of practical importance and practical informatics needs a solid foundation to develop its full potential.

Therefore, two thirds of the book are devoted to formal languages and automata theory. Formal languages are indispensable for applied computer science, since one meets them everywhere. Thus, we cover grammars (formalizing the generation), automata (formalizing the acceptance) and their interaction for regular and context-free languages.

The remaining third formalizes the intuitive notion of algorithm by introducing partial recursive functions and Turing machines. We show the equivalence of these two models and prove the existence of a universal Turing machine. That is, there is *one* computing device that can perform *every* possible computation. Finally, we show that there are problems which cannot be solved at all by any computer. Here we start with the halting problem, continue with Post's correspondence problem, and apply the theory developed to obtain a rather complete picture concerning problems arising naturally in formal language theory. Finally, we show the fixed point and recursion theorem, Rice's famous theorem, and introduce (abstract) complexity classes and study some of their properties.

The book also contains important material which is rarely covered in other textbooks, e.g., the Theorem of Chomsky-Schützenberger, our exposition of the partial recursive functions, Dedekind's justification theorem, abstract complexity classes and their properties.

Note that the book is accompanied by 15 sets of slides (in Japanese and English) which can be directly used in classrooms. We have tested these teaching materials at Hokkaido University. These slides and the book itself aim at Japanese students who wish to improve their English languages skills while studying their favorite subject and at international students who wish to complete their computer science studies at a Japanese university.

Experience shows that it is often difficult to find the correct technical term when dealing with a highly specialized subject such as theory of computation. We wish to facilitate this task and have included translations (English-Japanese) in the main text. If an important technical term is introduced, it is *emphasized* and followed by the appropriate Japanese term.

Definitions are written in italics except the term that is defined which is typeset in roman. Again, the term defined is translated into Japanese.

Of course, a textbook like the present one is also used as a reference. Consequently, we have added three indices, the usual subject index, a symbol index and a Japanese index to facilitate the search in such cases.

This book, like any other book has a certain disadvantage when compared with a teacher, i.e., it cannot speak and answer to questions. As a consequence, we have included exercises and problems. There is no principal difference between them such as the degree of difficulty. The only distinction is that the solutions to problems are *included* and the answers to exercises are *excluded*. These exercises and problems are recommended for classroom seminars and/or homework assignments, since they hopefully shed additional light on many interesting features. It should also be noted that our solutions are not the only possible ones. Sometimes, we included different solutions to the same problem to highlight this aspect.

Last but not least we confess that the course is demanding. But this is just in line with William S. Clark's (the founding vice-president of Hokkaido University) encouragement

```
                              Boys, be ambitious !
```
Of course, nowadays, we reformulate this encouragement as
```
                         Girls and Boys, be ambitious !
```

February, 2009

Thomas Zeugmann
Shin-ichi Minato
Yoshiaki Okubo

Contents

1. Introducing Formal Languages

1.1	Introduction	1
1.2	Basic Definitions and Notations	2
1.3	Strings and Languages	7
1.4	Palindromes	9
1.5	Problem Set 1	13

2. Introducing Formal Grammars

2.1	Defining Formal Grammars	14
2.2	Regular Languages	16
2.3	Problem Set 2	20

3. Finite State Automata

3.1	Computing with Finite Automata	22
3.2	Finite Automata and Regular Languages	25
3.3	Problem Set 3	30

4. Characterizations of $\mathcal{REG}$

4.1	Nerode's Theorem	31
4.2	Regular Expressions	33
4.3	The Pumping Lemma for Regular Languages	37
4.4	Problem Set 4	40

5. Regular Expressions in UNIX

5.1	Lexical Analysis	43
5.2	Finding Patterns in Text	45
5.3	Problem Set 5	47

6. Context-Free Languages

- 6.1 Defining Context-Free Languages ... *48*
- 6.2 Closure Properties for Context-Free Languages ... *49*
- 6.3 Problem Set 6 ... *58*

7. More About Context-Free Languages

- 7.1 Backus-Naur Form ... *59*
- 7.2 Parse Trees, Ambiguity ... *61*
- 7.3 Chomsky Normal Form ... *69*
- 7.4 Problem Set 7 ... *75*

8. $\mathcal{CF}$ and Homomorphisms

- 8.1 Substitutions and Homomorphisms ... *76*
- 8.2 Homomorphic Characterization of $\mathcal{CF}$ ... *82*
- 8.3 Problem Set 8 ... *91*

9. Pushdown Automata

- 9.1 Introducing Pushdown Automata ... *92*
- 9.2 PDAs and Context-Free Languages ... *100*
- 9.3 Problem Set 9 ... *104*

10. $\mathcal{CF}$, PDAs and Beyond

- 10.1 Greibach Normal Form ... *105*
- 10.2 Main Theorem ... *111*
- 10.3 Context-Sensitive Languages ... *114*
- 10.4 Problem Set 10 ... *117*

11. Models of Computation

- 11.1 Partial Recursive Functions ... *121*
- 11.2 Pairing Functions ... *131*
- 11.3 General Recursive Functions ... *133*

11.4 Problem Set 11 .... *136*

12. Turing Machines

12.1 One-tape Turing Machines *137*

12.2 Turing Computations *139*

12.3 The Universal Turing Machine *146*

12.4 Accepting Languages *150*

12.5 Problem Set 12 .... *152*

13. Algorithmic Unsolvability

13.1 The Halting Problem *153*

13.2 Post's Correspondence Problem *156*

13.3 Problem Set 13 .... *167*

14. Applications of PCP

14.1 Results for Context-Free Languages *168*

14.2 Back to Regular Languages *174*

14.3 Results concerning $\mathcal{L}_0$ *175*

14.4 Summary *180*

14.5 Problem Set 14 .... *181*

15. Numberings, Complexity

15.1 Gödel Numberings *183*

15.2 The Fixed Point, the Recursion, and Rice's Theorems ... *184*

15.3 Complexity *188*

15.4 Complexity Classes *195*

    15.4.1 Recursive Enumerability of Complexity Classes ... *197*

    15.4.2 An Undecidability Result *202*

    15.4.3 The Gap-Theorem *204*

15.5 Problem Set 15 .... *206*

Appendix ... *207*
A.1 Greek Alphabet （ギリシャ文字） ... *207*
Bibliography ... *208*
Subject Index ... *211*
和英索引 ... *217*
List of Symbols ... *221*

16. Solutions of Problems

Chapter 16 is on the CD-ROM, see chapter16.pdf

16.1 Solving Problem Set 1 ... *223*
16.2 Solving Problem Set 2 ... *226*
16.3 Solving Problem Set 3 ... *231*
16.4 Solving Problem Set 4 ... *237*
16.5 Solving Problem Set 5 ... *241*
16.6 Solving Problem Set 6 ... *247*
16.7 Solving Problem Set 7 ... *254*
16.8 Solving Problem Set 8 ... *257*
16.9 Solving Problem Set 9 ... *261*
16.10 Solving Problem Set 10 ... *265*
16.11 Solving Problem Set 11 ... *270*
16.12 Solving Problem Set 12 ... *275*
16.13 Solving Problem Set 13 ... *278*
16.14 Solving Problem Set 14 ... *282*
16.15 Solving Problem Set 15 ... *285*

付録 CD-ROM について

本書には，半期 (15 回) の講義で用いることを想定したスライド (日本語版・英語版)，および，すべての課題 (problem) の解答を収めた CD-ROM が付属しています．これらを使用するには，PDF ファイルを閲覧可能なソフトウェアのインストールが必要です．詳細については，CD-ROM 中の Readme.txt(英語)，J-Readme-SJIS.txt(日本語 Shift-JIS) もしくは J-Readme-UTF8.txt(日本語 UTF-8) を御覧下さい．

なお，使用にあたっては，以下の点に留意して下さい．

- 本ソフトウェアを商用で使用することはできません．
- 本ソフトウェアのコピーを他に流布することはできません．
- 本ソフトウェアの使用により生じた損害等については，著者ならびにコロナ社は一切の責任を負いません．

本 CD-ROM に関する不具合やお気付きの点については，コロナ社宛に御連絡下さい．

About the CD-ROM

The attached CD-ROM contains all solutions to the problems and 15 sets of lecture slides in Japanese and English, respectively, for one semester. All materials are provided in the PDF format. To enjoy the materials, you may need to install some PDF viewer. More about the CD-ROM can be found in Readme.txt (in English), J-Readme-SJIS.txt (in Japanese Shift-JIS) or J-Readme-UTF8.txt (in Japanese UTF-8) on the CD-ROM.

The CD-ROM must be used under the following restrictions:

- You may not use the CD-ROM for commercial use.
- You may not distribute copies of the CD-ROM by any means.
- The authors and the publisher are not responsible for the consequences of use of the CD-ROM.

Please send your comments and suggestions regarding the CD-ROM to the publisher.

本書の姉妹編「英語で学ぶ計算量と暗号理論」内容梗概

現代の暗号理論は計算量理論を基礎としている．そして，暗号および暗号解読が，計算量理論の最近の研究トピックの動機となっている．そこで本書では，暗号と暗号解読の入門解説に必要な計算量理論の側面に焦点をあてる．具体的には，べき乗余，素数判定，離散根数といった数論問題の計算量を取り上げる．さらに，基本的な計算量クラスをおさえ，完全問題を学んだ後，確率的計算量クラスについてもふれる．暗号理論では，その歴史を概観した後，公開鍵暗号の詳細を取り上げ，最後に，暗号プロトコルについて学ぶ．

なお，本書が対象とする読者は，計算機科学，数学，工学を学ぶ全ての大学生・大学院生であり，計算量と暗号理論，およびそれらの関係に興味を持つ者を想定している．

Outline of "Complexity and Cryptography" a companion volume

Modern cryptography is based on complexity theory and current topics studied in complexity theory derive their motivation to a certain extent from cryptography and cryptanalysis. Therefore, this book focuses on those aspects of complexity theory which we need for our introductory exposition of cryptography and cryptanalysis. We cover the complexity of number theoretic problems including modular exponentiation, primality testing and taking discrete roots. Then we study well-known complexity classes, look at complete problems and touch probabilistic complexity classes. Cryptography starts with a short historical sketch. Then public-key cryptography is covered in some more detail. Finally, we study cryptographic protocols.

The intended audience are all undergraduate and graduate students of computer science, mathematics, and engineering, who are interested in an introductory course to complexity and cryptography and their interrelation.

1 Introducing Formal Languages

1.1 Introduction

This book is about the study of a fascinating and important subject: *the theory of computation* (計算理論). It comprises the fundamental mathematical properties of computer hardware, software, and certain applications thereof. We are going to determine what can and cannot be computed. If it can, we also seek to figure out on which type of computational model, how quickly, and with how much memory.

Theory of computation has many connections with engineering practice, and, as a true science, it also comprises philosophical aspects.

Since formal languages are of fundamental importance to computer science, we shall start our course by having a closer look at them.

First, we clarify the subject of *formal language theory* (形式言語理論). Generally speaking, formal language theory concerns itself with sets of strings called *languages* and different mechanisms for generating and recognizing them. Mechanisms for generating sets of strings are usually referred to as *grammars* and mechanisms for recognizing sets of strings are called *acceptors* or *automata*. If we compare formal languages with natural languages, then mechanisms for generating sets of strings are needed to model the speaker and mechanisms for recognizing or accepting strings model the listener. Clearly, the speaker is supposed to generate exclusively sentences belonging to the language on hand. On the other hand, the listener has first to check if the sentence she is listening to does really belong to the language on hand before she can start to further reflect about its semantics. Recent research in neuro-biology has shown that humans indeed first parse sentences they are listening to before they start thinking about them. As a matter of fact, though the parsing process in the brain is not yet understood, it could be shown that parsing is usually done quite fast.

The same objective is sought for formal languages. That is, we wish to develop a theory such that generators and acceptors do coincide with respect to the set of strings generated and accepted, respectively.

A mathematical theory for generating and accepting languages emerged in the later 1950's and has been extensively developed since then. Nowadays there are elaborated theories for both computer languages and natural languages. Clearly, a one semester course is too short to cover not even the most beautiful parts. So we have to restrict ourselves to the most fundamental parts of formal language theory, i.e., to the regular languages, the context-free languages, and the recursively enumerable languages. This will suffice to obtain a basic understanding of what formal language theory is all about and what are the fundamental proof techniques. For having a common ground, we shortly recall the mathematical background needed.

1.2 Basic Definitions and Notations

For any set M, by card(M) and $\wp$(M) we denote the *cardinality* (要素数) of M and the *power set* (べき集合) of M, respectively. Let X, Y be any two

1.2 Basic Definitions and Notations

sets; then we use $X \cup Y$, $X \cap Y$ and $X \setminus Y$ to denote the union, intersection and difference of X and Y, respectively. By $\mathbb{N} = \{0, 1, 2, \ldots\}$ we denote the *set of all natural numbers*. We set $\mathbb{N}^+ = \mathbb{N} \setminus \{0\}$. By $\emptyset$ we denote the *empty set*. If we have countably many sets $X_0, X_1, \ldots$, then we use $\bigcup_{i \in \mathbb{N}} X_i$ to denote the union of all X_i, i.e.,

$$\bigcup_{i \in \mathbb{N}} X_i = X_0 \cup X_1 \cup \cdots \cup X_n \cup \cdots . \tag{1.1}$$

Analogously, we write $\bigcap_{i \in \mathbb{N}} X_i$ to denote the intersection of all X_i, i.e.,

$$\bigcap_{i \in \mathbb{N}} X_i = X_0 \cap X_1 \cap \cdots \cap X_n \cap \cdots . \tag{1.2}$$

It is useful to have the following notions. Let X, Y be any sets and let $f \colon X \to Y$ be a function. For any $y \in Y$ we define

$$f^{-1}(y) = \{x \mid x \in X, f(x) = y\} . \tag{1.3}$$

We refer to $f^{-1}(y)$ as the set of *pre-images* (原像) of y.

Also, we need the following definition.

Definition 1.1 *Let X, Y be any sets and let $f \colon X \to Y$ be a function. The function f is said to be*

(1) injective (単射) *if $f(x) = f(y)$ implies $x = y$ for all $x, y \in X$;*
(2) surjective (全射) *if for every $y \in Y$ there is an $x \in X$ such that $f(x) = y$;*
(3) bijective (全単射) *if f is injective and surjective.*

The following exercise relates Definition 1.1 to (1.3).

Exercise 1.1 *Let X, Y be any sets and let $f \colon X \to Y$ be a function. Then*

(1) *f is injective if $\operatorname{card}(f^{-1}(y)) \leq 1$ for all $y \in Y$;*
(2) *f is surjective if $\operatorname{card}(f^{-1}(y)) \geq 1$ for all $y \in Y$; and*
(3) *f is bijective if $\operatorname{card}(f^{-1}(y)) = 1$ for all $y \in Y$.*

Next, let X and Y be any sets. We say that X and Y have the same cardinality if there exists a bijection $f\colon X \to Y$. If a set X has the same cardinality as the set $\mathbb{N}$ of natural numbers, then we say that X is *countably infinite* (可算無限).

A set X is *at most countably infinite* (高々可算無限) if it is finite or countably infinite. If $X \neq \emptyset$ and X is at most countably infinite, then there exists a surjection $f\colon \mathbb{N} \to X$, i.e.,

$$X = \{f(0), f(1), f(2), \ldots\}.$$

So, intuitively, we can enumerate all the elements of X (where repetitions are allowed).

The following theorem is of fundamental importance.

Theorem 1.1 (Cantor's Theorem (カントールの定理)**)** *For every countably infinite set X the set $\wp(X)$ is not countably infinite.*

Proof. Since X is countably infinite, there is a bijection $f\colon \mathbb{N} \to X$. Suppose that $\wp(X)$ is countably infinite. Then there must exist a bijection $g\colon \mathbb{N} \to \wp(X)$. Now, we define a diagonal set D as follows:

$$D = \{f(j) \mid j \in \mathbb{N},\ f(j) \notin g(j)\}.$$

By construction, $D \subseteq X$ and therefore $D \in \wp(X)$. Consequently, there must be a number $d \in \mathbb{N}$ such that $D = g(d)$.

Now, we consider $f(d)$. By the definition of f we know that $f(d) \in X$. Since $D \subseteq X$, there are two possible cases, either $f(d) \in D$ or $f(d) \notin D$.

Case 1. $f(d) \in D$.

By the definition of D and d we directly get

$$f(d) \in D \iff f(d) \notin g(d) \iff f(d) \notin D,$$

since $g(d) = D$. This contradiction (矛盾) shows that Case 1 cannot happen.

1.2 Basic Definitions and Notations

Case 2. $f(d) \notin D$.

By construction, $f(d) \notin D$ holds if and only if $f(d) \in g(d)$ if and only if $f(d) \in D$, again a contradiction. Thus, Case 2 cannot happen either, and hence the supposition that $\wp(X)$ is countably infinite, cannot hold. ∎

Next, we recall the definition of binary relation. Let X, Y be any non-empty sets. We set $X \times Y = \{(x, y) \mid x \in X \text{ and } y \in Y\}$. Every $\rho \subseteq X \times Y$ is said to be a *binary relation* (二項関係). We sometimes use the notation $x \rho y$ instead of writing $(x, y) \in \rho$. Of special importance is the case where $X = Y$. If $\rho \subseteq X \times X$ then we also say that ρ is a binary relation over X.

Definition 1.2 *Let $X \neq \emptyset$ be any set, and let ρ be any binary relation over X. The relation ρ is said to be*

(1) reflexive (反射的) *if* $(x, x) \in \rho$ *for all* $x \in X$;

(2) symmetric (対称的) *if* $(x, y) \in \rho$ *implies* $(y, x) \in \rho$ *for all* $x, y \in X$;

(3) transitive (推移的) *if* $(x, y) \in \rho$ *and* $(y, z) \in \rho$ *implies* $(x, z) \in \rho$ *for all* $x, y, z \in X$;

(4) antisymmetric (反対称的) *if* $(x, y) \in \rho$ *and* $(y, x) \in \rho$ *implies* $x = y$.

Any binary relation satisfying (1) *through* (3) *is called* equivalence relation (同値関係). *For any* $x \in X$, *we write* $[x]$ *to denote the* equivalence class (同値類) *generated by* x, *i.e.*, $[x] = \{y \mid y \in X \text{ and } (x, y) \in \rho\}$. *We set* $X/\rho = \{[x] \mid x \in X\}$.

Any relation satisfying (1), (3) *and* (4) *is called* partial order (半順序). *In this case, we also say that* (X, ρ) *is a* partially ordered set (半順序集合).

Examples and further properties are provided at the end of the chapter as problems, and their solutions are given in Chapter 16.

Definition 1.3 *Let $\rho \subseteq X \times Y$ and $\tau \subseteq Y \times Z$ be binary relations. The composition* (合成) *of ρ and τ is the binary relation $\zeta \subseteq X \times Z$ defined as:*

$\zeta = \rho\tau$

$\quad = \{(x,z) \mid$ there is a $y \in Y$ such that $(x,y) \in \rho$ and $(y,z) \in \tau\}$.

Now, let $X \neq \emptyset$ be any set; there is a special binary relation ρ^0 called *equality* (恒等), and defined as $\rho^0 = \{(x,x) \mid x \in X\}$. Moreover, let $\rho \subseteq X \times X$ be any binary relation. We inductively define $\rho^{i+1} = \rho^i \rho$ for each $i \in \mathbb{N}$.

Definition 1.4 *Let $X \neq \emptyset$ be any set, and ρ be any binary relation over X. The* reflexive-transitive closure *(反射的推移的閉包) of ρ is the binary relation* $\rho^* = \bigcup_{i \in \mathbb{N}} \rho^i$.

We illustrate Definition 1.4 by using the following example. Define

$$\rho = \{(x, x+1) \mid x \in \mathbb{N}\}. \tag{1.4}$$

Then, $\rho^0 = \{(x,x) \mid x \in \mathbb{N}\}$, and $\rho^1 = \rho$. Next we compute

$$\rho^2 = \rho\rho = \{(x,z) \mid x, z \in \mathbb{N} \quad \text{such that there is a } y \in \mathbb{N}$$
$$\text{with } (x,y) \in \rho \text{ and } (y,z) \in \rho\}.$$

By the definition of ρ (see (1.4) above), $(x,y) \in \rho$ implies $y = x + 1$, and $(x+1, z) \in \rho$ implies $z = x + 2$. Hence, $\rho^2 = \{(x, x+2) \mid x \in \mathbb{N}\}$.

We proceed inductively. Taking into account that we have just proved the induction basis, we can assume the following induction hypothesis

$$\rho^i = \{(x, x+i) \mid x \in \mathbb{N}\}. \tag{1.5}$$

Claim. $\rho^{i+1} = \{(x, x+i+1) \mid x \in \mathbb{N}\}$.

By definition, $\rho^{i+1} = \rho^i \rho$, and thus, by the definition of composition and the induction hypothesis (see (1.5)) we get:

$$\rho^i \rho = \{(x,z) \mid x, z \in \mathbb{N}, \text{ there is a } y \text{ with } (x,y) \in \rho^i \text{ and } (y,z) \in \rho\}$$
$$= \{(x, x+i+1) \mid x \in \mathbb{N}\},$$

since $(x,y) \in \rho^i$ implies $y = x + i$. This proves the claim.

Finally, $\rho^* = \bigcup_{i \geq 0} \rho^i$, and therefore ρ^* is just the well-known binary relation "$\leq$" over $\mathbb{N}$, i.e., $(x, y) \in \rho^*$ if and only if $x \leq y$.

1.3 Strings and Languages

A formalism is required to deal with strings and sets of strings, and we therefore introduce it here. By Σ we denote a finite non-empty set called *alphabet* (アルファベット). The elements of Σ are assumed to be *indivisible symbols* (不可分記号) and referred to as *letters* (文字) or *symbols* (記号).

For example, $\Sigma = \{0, 1\}$ is an alphabet containing the letters 0 and 1, and $\Sigma = \{a, b, c\}$ is an alphabet containing the letters a, b, and c. In certain applications, e.g., in compiling, we may also have alphabets containing for example **begin** and **end**. But the letters **begin** and **end** are also assumed to be indivisible.

Definition 1.5 *A* string (文字列) *over an alphabet Σ is a finite length sequence of letters from Σ. A typical string is written as* $s = a_1 a_2 \cdots a_k$, *where $a_i \in \Sigma$ for $i = 1, \ldots, k$.*

Note that we also allow $k = 0$ resulting in the *empty string* (空文字列) which we denote by λ. We call k the *length* (長さ) of s and denote it by $|s|$, so $|\lambda| = 0$. By Σ^* we denote the set of all strings over Σ, and we set $\Sigma^+ = \Sigma^* \setminus \{\lambda\}$. Now, let $s, w \in \Sigma^*$; we define a binary operation called *concatenation* (連接) (or word product). The concatenation of s and w is the string sw. For example, let $\Sigma = \{0, 1\}$, $s = 000111$ and $w = 0011$; then $sw = 0001110011$.

Proposition 1.1 summarizes the basic properties of concatenation[†].

[†] Because of these properties, Σ^* is also referred to as *free monoid* (自由モノイド) in the literature.

8   1. Introducing Formal Languages

Proposition 1.1 *Let Σ be any alphabet.*
(1) *Concatenation is associative (結合的), i.e., for all $x, y, z \in \Sigma^*$, we have* $x(yz) = (xy)z$.
(2) *The empty string λ is a two-sided identity (両側単位元) for Σ^*, i.e., for all $x \in \Sigma^*$, we have* $x\lambda = \lambda x = x$.
(3) Σ^* *is free of nontrivial identities, i.e., for all $x, y, z \in \Sigma^*$,*
   i) $zx = zy$ *implies* $x = y$ *; and*
   ii) $xz = yz$ *implies* $x = y$.
(4) *For all $x, y \in \Sigma^*$, $|xy| = |x| + |y|$.*

Next, we extend our operations on strings to sets of strings. Let X, Y be sets of strings. Then the *product* (直積) of X and Y is defined as

$$XY = \{xy \mid x \in X \text{ and } y \in Y\}. \tag{1.6}$$

Let $X \subseteq \Sigma^*$; define $X^0 = \{\lambda\}$ and for all $i \geq 0$ set $X^{i+1} = X^i X$. The *Kleene closure* (クリーニ閉包) of X is defined as $X^* = \bigcup_{i \in \mathbb{N}} X^i$, and the *semigroup closure* (半群閉包) of X is $X^+ = \bigcup_{i \in \mathbb{N}^+} X^i$.

Finally, we define the *transpose* of a string and of sets of strings.

Definition 1.6 *Let Σ be any alphabet. The* transpose (転置、反転) *operator is defined on strings in Σ^* as follows:*

$$\lambda^T = \lambda, \text{ and}$$
$$(xa)^T = a(x^T) \quad \text{for all } x \in \Sigma^* \text{ and all } a \in \Sigma.$$

We extend it to sets X of strings by setting $X^T = \{x^T \mid x \in X\}$. For example, let $s = abbcc$, then $s^T = ccbba$. Furthermore, let $X = \{a^i b^j \mid i, j \in \mathbb{N}^+\}$, then $X^T = \{b^j a^i \mid i, j \in \mathbb{N}^+\}$. Here, a^i is defined by setting $a^0 = \lambda$ and $a^{i+1} = a^i a$ for all $i \in \mathbb{N}$.

We continue by defining languages.

Definition 1.7 *Let Σ be any alphabet. Every subset $L \subseteq \Sigma^*$ is called* language (言語).

Note that the empty set as well as $L = \{\lambda\}$ are also languages. Next, we ask how many languages there are. Let $m = \text{card}(\Sigma)$. There is precisely one string of length 0, i.e., λ, there are m strings of length 1, i.e., a for all $a \in \Sigma$, there are m^2 many strings of length 2, and in general there are m^n many strings of length n. Thus, the cardinality of Σ^* is countably infinite. Therefore, by Theorem 1.1 we conclude that there are *uncountably* (非可算個) many languages (as much as there are real numbers). Since the generation and recognition of languages should be done algorithmically, we immediately see that only countably many languages can be generated and recognized by an algorithm.

1.4 Palindromes

Next, we look at something interesting from natural languages and ask how we can put this into the framework developed so far. That is, we want to look at palindromes. A *palindrome* (回文) is a string that reads the same from left to right and from right to left. Consider the following strings.

トビコミコビト

にわとりとことりとわに

テングノハハノグンテ

AKASAKA, or removing space and punctuation symbols, the famous self-introduction of Adam to Eve: madamimadam (Madam, I'm Adam).

Now we ask how can we describe the language of all palindromes over the alphabet $\{a, b\}$ (just to keep it simple). We may have seen inductive definitions in arithmetic, e.g., of the faculty function defined over $\mathbb{N}$ and denoted $n!$. It can be inductively defined as $0! = 1$ and $(n+1)! = (n+1)n!$.

One of the nice properties of free monoids is that we can adopt the concepts of "inductive definition" and "proof by induction."

So let us try it. Of course λ, a, and b are palindromes. Every palindrome must begin and end with the same letter, and if we remove the first and last letter of a palindrome, we still get a palindrome. This observation suggests the following basis and induction for defining L_{pal}.

Induction Basis: λ, a, and b are palindromes.

Induction Step: If $w \in \{a, b\}^*$ is a palindrome, then awa and bwb are also palindromes. Furthermore, no string $w \in \{a, b\}^*$ is a palindrome, unless it follows from this basis and induction step.

But stop, we could have also used the transpose operator T to define the language of all palindromes, i.e., $\tilde{L}_{pal} = \{w \in \{a, b\}^* \mid w = w^T\}$.

We used a different notation in the latter definition, since we still do not know whether or not $L_{pal} = \tilde{L}_{pal}$. For getting this equality, we need a proof.

Theorem 1.2 $L_{pal} = \tilde{L}_{pal}$.

Proof. Equality of sets X, Y is often proved by showing $X \subseteq Y$ and $Y \subseteq X$. So, let us first show that $L_{pal} \subseteq \tilde{L}_{pal}$.

We start with the strings defined by the basis, i.e., λ, a, and b. By the definition of the transpose operator, we have $\lambda^T = \lambda$. Thus, $\lambda \in \tilde{L}_{pal}$. Next, we deal with a. In order to apply the definition of the transpose operator, we use Property (2) of Proposition 1.1, i.e., $a = \lambda a$. Then, we have

$$a^T = (\lambda a)^T = a\lambda^T = a\lambda = a .$$

The proof for b is analogous and thus omitted.

Now, the induction hypothesis is that for all strings w with $|w| \leq n$, we have $w \in L_{pal}$ implies $w \in \tilde{L}_{pal}$. In accordance with our definition of L_{pal}, the induction step is from n to $n + 2$. Let $w \in L_{pal}$ be any string with

1.4 Palindromes

$|w| = n+2$. Thus, $w = ava$ or $w = bvb$, where $v \in \{a,b\}^*$ such that $|v| = n$. Then v is a palindrome in the sense of the definition of L_{pal}, and by the induction hypothesis, we obtain $v = v^T$. Now, we have to establish the following claims providing a special property of the transpose operator.

Claim 1. Let Σ be any alphabet, $n \in \mathbb{N}^+$, and $w = w_1 \ldots w_n \in \Sigma^*$, where $w_i \in \Sigma$ for all $i \in \{1, \ldots, n\}$. Then $w^T = w_n \ldots w_1$.

The proof is by induction. The induction basis is for $w_1 \in \Sigma$ and done as above. Now, we have the induction hypothesis that $(w_1 \ldots w_n)^T = w_n \ldots w_1$. The induction step is from n to $n+1$ and done as follows.

$$(w_1 \ldots w_n w_{n+1})^T = w_{n+1}(w_1 \ldots w_n)^T = w_{n+1} w_n \ldots w_1 \,.$$

Note that the first equality above is by the definition of the transpose operator and the second one by the induction hypothesis. Thus, Claim 1 is proved.

Claim 2. For all $n \in \mathbb{N}$, if $p = p_1 x p_{n+2}$ then $p^T = p_{n+2} x^T p_1$ for all $p_1, p_{n+2} \in \{a,b\}$ and $x \in \{a,b\}^*$, where $|x| = n$.

Let $p = p_1 x p_{n+2}$ and let $x = x_1 \ldots x_n$, where $x_i \in \Sigma$. Then, $p = p_1 x_1 \ldots x_n p_{n+2}$ and by Claim 1, we have $p^T = p_{n+2} x_n \ldots x_1 p_1$ as well as $x^T = x_n \ldots x_1$. Hence, $p_{n+2} x^T p_1 = p^T$ (see Proposition 1.1) and Claim 2 is shown.

Consequently, by using Claim 2 just established we get

$$w^T = (ava)^T = av^T a \underbrace{=}_{\text{by IH}} ava = w \,.$$

Again, the case $w = bvb$ can be handled analogously and is thus omitted.

For completing the proof, we have to show $\tilde{L}_{pal} \subseteq L_{pal}$. For the induction basis, we know that $\lambda = \lambda^T$, i.e., $\lambda \in \tilde{L}_{pal}$ and by the induction basis of the definition of L_{pal}, we also know that $\lambda \in L_{pal}$.

Thus, we have the induction hypothesis that for all strings w of length n: if $w = w^T$ then $w \in L_{pal}$.

The induction step is from n to $n+1$. That is, we have to show: if $|w| = n+1$ and $w = w^T$ then $w \in L_{pal}$.

Since the case $n = 1$ directly results in a and b and since $a, b \in L_{pal}$, we assume $n > 1$ in the following. So, let $w \in \{a, b\}$ be any string with $|w| = n+1$ and $w = w^T$, say $w = a_1 \ldots a_{n+1}$, where $a_i \in \Sigma$. Thus, by assumption we have $a_1 \ldots a_{n+1} = a_{n+1} \ldots a_1$.

Now, applying Property (3) of Proposition 1.1 directly yields $a_1 = a_{n+1}$. We have to distinguish the cases $a_1 = a$ and $a_1 = b$. Since both cases can be handled analogously, we consider only the case $a_1 = a$ here. Thus, we can conclude $w = ava$, where $v \in \{a, b\}^*$ and $|v| = n-1$. Next, applying the property of the transpose operator established above, we obtain $v = v^T$, i.e., $v \in L_{pal}$. Finally, the "induction" part of the definition of L_{pal} directly implies $w \in L_{pal}$. ∎

Since we shall see the language L_{pal} and variations of it in this book occasionally, please ensure that you have understood what we have done above.

We finish this chapter by recommending to solve the following exercises and problems.

Exercise 1.2 *Consider the following relation $\rho \subseteq \mathbb{N} \times \mathbb{N}$ defined as*

$$\rho = \{(x+1, x) \mid x \in \mathbb{N}\}.$$

Compute the reflexive-transitive closure of ρ.

Exercise 1.3 *Determine which of the following sets Σ are alphabets:*

(1) $\Sigma = \mathbb{N}$,

(2) $\Sigma = \{0, 01, 10, 1\}$,

(3) $\Sigma = \{A, B, C, S, T\}$.

Exercise 1.4 *Let Σ be any alphabet and let $X \subseteq \Sigma^*$ be any set. Which of the following statements is true?*

(a) $\lambda \in X^*$,

(b) $\lambda \in X^+$.

1.5 Problem Set 1

Problem 1.1 Let $X = \mathbb{N}$. Define $x\rho y$ if x divides y. Prove or disprove (証明または反証せよ):

(a) $(\mathbb{N}, \rho)$ is a partially ordered set.

(b) The relation ρ is an equivalence relation.

Problem 1.2 Let X be any non-empty set and let ρ be any equivalence relation over X. Prove or disprove that the set of all equivalence classes forms a partition of the set X, i.e., for all $x, y \in X$ we have:

(a) either $[x] = [y]$ or $[x] \cap [y] = \emptyset$;

(b) $\bigcup_{x \in X} [x] = X$.

Problem 1.3 Prove or disprove: For every binary relation ρ over a set X we have $\rho^* = (\rho^*)^*$, i.e., the reflexive–transitive closure of the reflexive–transitive closure is the reflexive–transitive closure itself.

2 Introducing Formal Grammars

We have to formalize what is meant by generating a language. If we look at natural languages, then we have the following situation. The set Σ consists of all words in the language. Although large, Σ is finite. What is usually done in speaking or writing natural languages is forming sentences. A typical sentence starts with a noun phrase followed by a verb phrase. Thus, we may describe this generation by

$$< \text{sentence} > \;\rightarrow\; < \text{noun phrase} >< \text{verb phrase} > .$$

Clearly, more complicated sentences are generated by more complicated rules. If we look in a usual grammar book, e.g., for the English language, then we see that there are, however, only finitely many rules for generating sentences. This suggest the following general definition of a grammar.

2.1 Defining Formal Grammars

Definition 2.1 *A 4-tuple* $\mathcal{G} = [T, N, \sigma, P]$ *is called a* grammar (文法) *if*
(1) T *and* N *are alphabets with* $T \cap N = \emptyset$,
(2) $\sigma \in N$,
(3) $P \subseteq ((T \cup N)^+ \setminus T^*) \times (T \cup N)^*$ *is finite.*

We call T the *terminal alphabet* (終端アルファベット), N the *nonterminal alphabet* (非終端アルファベット), σ the *start symbol* (開始記号) and P the set of *productions* (生成規則[†]) (or *rules* (規則)). Usually, productions are written in the form $\alpha \to \beta$, where $\alpha \in (T \cup N)^+ \setminus T^*$ and $\beta \in (T \cup N)^*$.

Next, we have to explain how to generate a language using a grammar. This is done by the following definition.

Definition 2.2 *Let* $\mathcal{G} = [T, N, \sigma, P]$ *be a grammar. Let* $\alpha', \beta' \in (T \cup N)^*$. *String* α' *is said to* directly generate (直接生成する) β', *written* $\alpha' \Rightarrow \beta'$, *if there are* $\alpha_1, \alpha_2, \alpha, \beta \in (T \cup N)^*$ *such that* $\alpha' = \alpha_1 \alpha \alpha_2$, $\beta' = \alpha_1 \beta \alpha_2$ *and* $\alpha \to \beta \in P$. *We write* $\stackrel{*}{\Rightarrow}$ *for the reflexive-transitive closure of* $\Rightarrow$.

Example 2.1 Let $\mathcal{G} = [\{a, b\}, \{\sigma\}, \sigma, P]$, where
$P = \{\sigma \to \lambda, \sigma \to a, \sigma \to b, \sigma \to a\sigma a, \sigma \to b\sigma b\}$.
Then we can directly generate a from σ, since $\sigma \to a$ is in P. Furthermore, we can generate the string $abba$ from σ as follows by using the rules $\sigma \to a\sigma a, \sigma \to b\sigma b$ and $\sigma \to \lambda$, i.e., we obtain

$$\sigma \Rightarrow a\sigma a \Rightarrow ab\sigma ba \Rightarrow abba . \tag{2.1}$$

A sequence like (2.1) is called a *generation* (生成) or *derivation* (導出). If a string s can be generated from a nonterminal h then we write $h \stackrel{*}{\Rightarrow} s$.

Finally, we can define the language generated by a grammar.

Definition 2.3 *Let* $\mathcal{G} = [T, N, \sigma, P]$ *be a grammar. The* language $L(\mathcal{G})$ *generated by* $\mathcal{G}$ *is defined as* $L(\mathcal{G}) = \{s \mid s \in T^* \text{ and } \sigma \stackrel{*}{\Rightarrow} s\}$.

The family of all languages that can be generated by a grammar in the sense of Definition 2.1 is denoted by $\mathcal{L}_0$. These languages are also called *type-0 languages* (0 型言語), where 0 should remind us to zero restrictions.

Example 2.2 We look again at the languages of all palindromes over the alphabet $\{a, b\}$, i.e., $L_{pal} = \{w \in \{a, b\}^* \mid w = w^T\}$. Let $\mathcal{G}$ be the grammar from Example 2.1. We claim that $L_{pal} = L(\mathcal{G})$. The proof is done by showing the following two claims inductively.

[†] 書き換え規則とも呼ぶ.

Claim 1. $L_{pal} \subseteq L(\mathcal{G})$.

Induction Basis: Consider $w = \lambda$, $w = a$ and $w = b$. Since P contains $\sigma \to \lambda$, $\sigma \to a$, and $\sigma \to b$, we get $\sigma \stackrel{*}{\Rightarrow} w$ in all three cases.

Induction Step: Now let $|w| \geq 2$. Since $w = w^T$, the string w must begin and end with the same symbol, i.e., $w = ava$ or $w = bvb$, where v must be a palindrome, too. By the induction hypothesis, we have $\sigma \stackrel{*}{\Rightarrow} v$, and thus

$\sigma \Rightarrow a\sigma a \stackrel{*}{\Rightarrow} ava$ proving the $w = ava$ case, or

$\sigma \Rightarrow b\sigma b \stackrel{*}{\Rightarrow} bvb$ proving the $w = bvb$ case.

This shows Claim 1.

Claim 2. $L(\mathcal{G}) \subseteq L_{pal}$.

Induction Basis: If the generation is done in one step, then one of the productions not containing σ on the right hand side must have been used, i.e., $\sigma \to \lambda$, $\sigma \to a$, or $\sigma \to b$. Thus, $\sigma \Rightarrow w$ results in $w = \lambda$, $w = a$ or $w = b$; hence $w \in L_{pal}$.

Induction Step: Suppose, the generation takes $n + 1$ steps, $n \geq 1$. Thus,

$\sigma \Rightarrow a\sigma a \stackrel{*}{\Rightarrow} ava$ or

$\sigma \Rightarrow b\sigma b \stackrel{*}{\Rightarrow} bvb$.

Since by the induction hypothesis, we know that $v \in L_{pal}$, we get in both cases $w \in L_{pal}$. ∎

Next, we study subclasses of grammars and the languages they generate. We start with the easiest subclass, the so-called regular languages.

2.2 Regular Languages

Definition 2.4 *A grammar* $\mathcal{G} = [T, N, \sigma, P]$ *is said to be* regular (正規文法) *provided for all* $\alpha \to \beta$ *in* P *we have* $\alpha \in N$ *and* $\beta \in T^* \cup T^*N$.

2.2 Regular Languages

Definition 2.5 *A language* L *is said to be* regular (正規言語) *if there exists a regular grammar* $\mathcal{G}$ *such that* $\mathsf{L} = \mathsf{L}(\mathcal{G})$. *By* $\mathcal{REG}$ *we denote the set of all regular languages.*

Example 2.3 Let $\mathcal{G} = [\{a, b\}, \{\sigma\}, \sigma, \mathsf{P}]$ with $\mathsf{P} = \{\sigma \to ab, \sigma \to a\sigma\}$. The grammar $\mathcal{G}$ is regular and $\mathsf{L}(\mathcal{G}) = \{a^n b \mid n \in \mathbb{N}^+\}$.

Example 2.4 Let $\mathcal{G} = [\{a, b\}, \{\sigma\}, \sigma, \mathsf{P}]$ with
$\mathsf{P} = \{\sigma \to \lambda,\ \sigma \to a\sigma,\ \sigma \to b\sigma\}$.
Again, $\mathcal{G}$ is regular and $\mathsf{L}(\mathcal{G}) = \Sigma^*$. So, Σ^* is a regular language.

Example 2.5 Let Σ be any alphabet, and let $\mathsf{X} \subseteq \Sigma^*$ be any finite set. Then, for $\mathcal{G} = [\Sigma, \{\sigma\}, \sigma, \mathsf{P}]$ with $\mathsf{P} = \{\sigma \to s \mid s \in \mathsf{X}\}$, we have $\mathsf{L}(\mathcal{G}) = \mathsf{X}$. Consequently, every *finite* language is regular.

Note, however, that the grammar given in Example 2.1 is *not* regular, since the productions $\sigma \to a\sigma a$ and $\sigma \to b\sigma b$ violate the condition of Definition 2.4.

Now, we have already seen several examples for regular languages. As curious as we are, we are going to ask which languages are regular. For answering this question, first, we deal with *closure properties* (閉包性).

Theorem 2.1 *The regular languages are closed under union, product and Kleene closure.*

Proof. Let L_1 and L_2 be any regular languages. Since $\mathsf{L}_1, \mathsf{L}_2 \in \mathcal{REG}$, there are regular grammars $\mathcal{G}_1 = [\mathsf{T}_1, \mathsf{N}_1, \sigma_1, \mathsf{P}_1]$ and $\mathcal{G}_2 = [\mathsf{T}_2, \mathsf{N}_2, \sigma_2, \mathsf{P}_2]$ such that $\mathsf{L}_i = \mathsf{L}(\mathcal{G}_i)$ for $i = 1, 2$. Without loss of generality, we may assume that $\mathsf{N}_1 \cap \mathsf{N}_2 = \emptyset$, for otherwise we simply rename the nonterminals appropriately.

Let us start with union. We have to show that $\mathsf{L} = \mathsf{L}_1 \cup \mathsf{L}_2$ is regular. The desired grammar $\mathcal{G}_{union}$ is defined as follows.

$$\mathcal{G}_{union} = [\mathsf{T}_1 \cup \mathsf{T}_2, \mathsf{N}_1 \cup \mathsf{N}_2 \cup \{\sigma\}, \sigma, \mathsf{P}_1 \cup \mathsf{P}_2 \cup \{\sigma \to \sigma_1, \sigma \to \sigma_2\}]\ .$$

By construction, $\mathcal{G}_{union}$ is regular.

Claim 1. $\mathsf{L} = \mathsf{L}(\mathcal{G}_{union})$.

We have to start every generation of strings with σ. Thus, there are two possibilities, i.e., $\sigma \rightarrow \sigma_1$ and $\sigma \rightarrow \sigma_2$. In the first case, we can continue with all generations that start with σ_1 yielding all strings in L_1. In the second case, we can continue with σ_2, thus getting all strings in L_2. Consequently, $\mathsf{L}_1 \cup \mathsf{L}_2 \subseteq \mathsf{L}$.

On the other hand, $\mathsf{L} \subseteq \mathsf{L}_1 \cup \mathsf{L}_2$ by construction. Hence, $\mathsf{L} = \mathsf{L}_1 \cup \mathsf{L}_2$.

We continue with the closure under product. So we have to show that $\mathsf{L}_1 \mathsf{L}_2$ is regular. A first idea might be to use a construction analogous to the one above, i.e., to take as a new starting production $\sigma \rightarrow \sigma_1 \sigma_2$. Unfortunately, this production is *not* regular. We have to be a bit more careful. But the underlying idea is fine, we just have to replace it by a sequential construction. The idea for doing that is easily described. Let $s_1 \in \mathsf{L}_1$ and $s_2 \in \mathsf{L}_2$. We want to generate $s_1 s_2$. Then, starting with σ_1 there is a generation $\sigma_1 \Rightarrow w_1 \Rightarrow w_2 \Rightarrow \cdots \Rightarrow s_1$. But instead of finishing the generation at this point, we want to have the possibility to continue to generate s_2. Thus, all we need is a production having a right hand side resulting in $s_1 \sigma_2$. This idea can be formalized as follows: Let

$$\mathcal{G}_{prod} = [\mathsf{T}_1 \cup \mathsf{T}_2, \mathsf{N}_1 \cup \mathsf{N}_2, \sigma_1, \mathsf{P}], \quad \text{where}$$
$$\mathsf{P} = \mathsf{P}_1 \setminus \{h \rightarrow s \mid s \in \mathsf{T}_1^* \text{ and } h \in \mathsf{N}_1\}$$
$$\cup \{h \rightarrow s\sigma_2 \mid h \rightarrow s \in \mathsf{P}_1 \text{ and } s \in \mathsf{T}_1^*\} \cup \mathsf{P}_2.$$

By its definition, $\mathcal{G}_{prod}$ is regular.

Claim 2. $\mathsf{L}(\mathcal{G}_{prod}) = \mathsf{L}_1 \mathsf{L}_2$.

By construction, we have $\mathsf{L}_1 \mathsf{L}_2 \subseteq \mathsf{L}(\mathcal{G}_{prod})$. For showing $\mathsf{L}(\mathcal{G}_{prod}) \subseteq \mathsf{L}_1 \mathsf{L}_2$, let $s \in \mathsf{L}_1 \mathsf{L}_2$. Hence, there are strings $s_1 \in \mathsf{L}_1$ and $s_2 \in \mathsf{L}_2$ such that $s = s_1 s_2$. Since $s_1 \in \mathsf{L}_1$, there is a generation $\sigma_1 \Rightarrow w_1 \Rightarrow \cdots \Rightarrow w_n \Rightarrow s_1$

2.2 Regular Languages

in $\mathcal{G}_1$. Note that w_n must contain precisely one nonterminal, say h, and it must be of the form $w_n = wh$, by Definition 2.4. Now, since $w_n \Rightarrow s_1$ and $s_1 \in T_1^*$, we must have applied a production $h \to s$ with $s \in T_1^*$ such that $wh \Rightarrow ws = s_1$. But in $\mathcal{G}_{prod}$ all these productions have been replaced by $h \to s\sigma_2$. Therefore, the last generation $w_n \Rightarrow s_1$ is now replaced by $wh \Rightarrow ws\sigma_2$. All what is left, is now applying the productions from P_2 to generate s_2 which is possible, since $s_2 \in L_2$. This proves the claim.

Finally, we have to deal with the Kleene closure. Let L be a regular language, and let $\mathcal{G} = [T, N, \sigma, P]$ be a regular grammar such that $L = L(\mathcal{G})$. We have to show that L^* is regular. By definition $L^* = \bigcup_{i \in \mathbb{N}} L^i$. Since $L^0 = \{\lambda\}$, we have to make sure that λ can be generated. This is obvious if $\lambda \in L$. Otherwise, we simply add the production $\sigma \to \lambda$. The rest is done analogously as in the product case, i.e., we set

$$\mathcal{G}^* = [T, N \cup \{\sigma^*\}, \sigma^*, P^*] , \quad \text{where}$$

$$P^* = P \cup \{h \to s\sigma \mid h \to s \in P \text{ and } s \in T^*\} \cup \{\sigma^* \to \sigma, \sigma^* \to \lambda\} .$$

We leave it as an exercise to prove $L(\mathcal{G}^*) = L^*$. ∎

Exercise 2.1 *Let* $L = \{a^n b^m \mid n, m \in \mathbb{N}\}$. *Prove that* $L \in \mathcal{REG}$.

Finally, we define the equivalence of grammars.

Definition 2.6 *Let* $\mathcal{G}$ *and* $\hat{\mathcal{G}}$ *be any grammars.* $\mathcal{G}$ *and* $\hat{\mathcal{G}}$ *are said to be equivalent (等価) if* $L(\mathcal{G}) = L(\hat{\mathcal{G}})$.

In order to have an example for equivalent grammars, we consider
$\mathcal{G} = [\{a\}, \{\sigma\}, \sigma, \{\sigma \to a\sigma a, \sigma \to aa, \sigma \to a\}]$,
and the following grammar
$\hat{\mathcal{G}} = [\{a\}, \{\sigma\}, \sigma, \{\sigma \to a, \sigma \to a\sigma\}]$.

Now, it is easy to see that $L(\mathcal{G}) = \{a\}^+ = L(\hat{\mathcal{G}})$, and hence $\mathcal{G}$ and $\hat{\mathcal{G}}$ are equivalent. Note however that $\hat{\mathcal{G}}$ is regular while $\mathcal{G}$ is not.

Exercise 2.2 *Provide a regular grammar $\mathcal{G}$ for the language*

$$L = \{abc, aabc, abacc\}.$$

Exercise 2.3 *Provide a regular grammar $\mathcal{G}$ for the language*

$$L = \{w \in \{a, b\}^* \mid 2 \text{ divides } |w|\},$$

and prove $L = L(\mathcal{G})$.

We finish this chapter by recommending to solve the following problems.

2.3 Problem Set 2

Problem 2.1 Construct a regular grammar which generates precisely all binary numbers that are divisible by 4. Prove the correctness of your grammar.

Problem 2.2 *Let $T = \{a, b, c\}$, $N = \{\sigma, \alpha, \beta\}$ and let $\mathcal{G} = [T, N, \sigma, P]$, where P is the set of the following productions:*

(1) $\sigma \to abc$ (5) $b\beta \to \beta b$

(2) $\sigma \to a\alpha bc$ (6) $a\beta \to aa\alpha$

(3) $\alpha b \to b\alpha$ (7) $a\beta \to aa$

(4) $\alpha c \to \beta bcc$

(1) Is $\mathcal{G}$ a regular grammar? Provide arguments justifying your answer.

(2) Determine $L(\mathcal{G})$.

(3) Prove the correctness of the assertion you made in (2).

Problem 2.3 Let $k \in \mathbb{N}^+$ be arbitrarily fixed. Construct a regular grammar for the language

$$L_k = \{w \in \{a, b\}^* \mid w = w_1 \ldots w_\ell,\ \ell > k,\ w_1 = \cdots = w_k = a$$
$$\text{and } w_{k+1} = \cdots = w_\ell = b\}.$$

Prove the correctness of your grammar.

3
Finite State Automata

In the previous chapter we learned how to formalize the generation of languages. This part looked at formal languages from the perspective of a speaker. Now, we turn our attention to accepting languages, i.e., we are going to formalize the perspective of a listener. In this chapter we deal with regular languages, and the machine model accepting them. The overall goal can be described as follows. Let Σ be again an alphabet, and let $L \subseteq \Sigma^*$ be any regular language. Now, for every string $s \in \Sigma^*$ we want to have a possibility to decide whether or not $s \in L$. Looking at the definition of a regular grammar, the following method may be easily discovered. We start generating strings until one of the following two conditions happens. First, the string s is generated. Clearly, then $s \in L$. Second, the length of our string s is exceeded. Now, taking into account that all further generable strings must be longer, we may conclude that $s \notin L$. There is only one problem with this method, i.e., its efficiency. It may take time that is exponential in the length of the input string s to terminate. Besides that, this approach hardly reflects what humans are doing when accepting sentences of natural languages. We therefore favor a different approach which we define next[†].

[†] Note that M.O. Rabin and D.S. Scott received the Turing Award in 1976 for their paper *Finite Automata and Their Decision Problems* [30], which introduced the idea of nondeterministic machines – a concept which has proved to be enormously valuable.

Definition 3.1 *A 5-tuple* $\mathcal{A} = [\Sigma, Q, \delta, q_0, F]$ *is said to be a* nondeterministic finite automaton (非決定性有限オートマトン) *if*

(1) Σ *is an alphabet (called the* input alphabet (入力アルファベット) *)*,

(2) Q *is a finite nonempty set (the* set of states (状態集合) *)*,

(3) $\delta\colon Q \times \Sigma \to \wp(Q)$, *the* transition relation (遷移関係),

(4) $q_0 \in Q$, *the* initial state (初期状態), *and*

(5) $F \subseteq Q$, *the* set of final states (最終状態集合).

There is also a deterministic counterpart of a nondeterministic finite automaton which we define next.

Definition 3.2 *A 5-tuple* $\mathcal{A} = [\Sigma, Q, \delta, q_0, F]$ *is said to be a* deterministic finite automaton (決定性有限オートマトン) *if*

(1) Σ *is an alphabet (called the* input alphabet (入力アルファベット) *)*,

(2) Q *is a finite nonempty set (the* set of states (状態集合) *)*,

(3) $\delta\colon Q \times \Sigma \to Q$, *the* transition function (遷移関数), *which must be defined for every input,*

(4) $q_0 \in Q$, *the* initial state (初期状態), *and*

(5) $F \subseteq Q$, *the* set of final states (最終状態集合).

If we do not want to specify whether an automaton is deterministic or nondeterministic, we refer to it as a *finite automaton* (有限オートマトン) . We use the abbreviation NFA and DFA to refer to a nondeterministic finite automaton and a deterministic finite automaton, respectively.

3.1 Computing with Finite Automata

So far, we have explained what a finite automaton is but not what it does. In order to explain how to compute with a finite automaton, we need some more definitions. For the deterministic case, we can easily define the

3.1 Computing with Finite Automata

language accepted by a finite automaton. First, we *extend* the definition of δ to strings. That is, we inductively define a function

$$\delta^*: Q \times \Sigma^* \to Q \,, \tag{3.1}$$

by setting

$$\delta^*(q, \lambda) = q \quad \text{for all } q \in Q \,,$$
$$\delta^*(q, sa) = \delta(\delta^*(q, s), a) \quad \text{for all } s \in \Sigma^*, \text{ all } a \in \Sigma, \text{ and all } q \in Q \,.$$

The proof of the following lemma is left as an exercise.

Lemma 3.1 *Let* $\mathcal{A} = [\Sigma, Q, \delta, q_0, F]$ *be a DFA. Then for all strings* $v, w \in \Sigma^*$ *and all* $q \in Q$ *we have* $\delta^*(q, vw) = \delta^*(\delta^*(q, v), w)$.

Definition 3.3 *Let* $\mathcal{A} = [\Sigma, Q, \delta, q_0, F]$ *be a DFA. The language* $L(\mathcal{A})$ *accepted* (受理される) *by* $\mathcal{A}$ *is* $L(\mathcal{A}) = \{s \mid s \in \Sigma^*, \delta^*(q_0, s) \in F\}$.

If we have $s \in L(\mathcal{A})$ for a string $s \in \Sigma^*$ then we say that there is an *accepting computation* (受理計算) for s. We adopt this notion also to NFAs. Note that $\lambda \in L(\mathcal{A})$ iff $q_0 \in F$.

In order to keep notation simple, in the following we shall identify δ^* with δ. It should be clear from context what is meant.

This seems very abstract, and some explanation is in order. Conceptually, a finite automaton possesses an *input tape* (入力テープ) that is divided into cells. Each cell can store a symbol from Σ or it may be empty. Moreover, a finite automaton has a *head* (ヘッド) to read what is stored in the cells. Initially, a string $s = s_1 s_2 \cdots s_k$ is written on the tape and the head is positioned on the leftmost symbol of the input, i.e., on s_1 (cf. Figure 3.1).

Moreover, the automaton is put into its initial state q_0. Now, the automaton reads s_1. Then, it changes its state to one of the possible states in $\delta(q_0, s_1)$, say q, and the head moves right to the next cell. Note that in the deterministic case, the state $\delta(q_0, s_1)$ is uniquely defined. Next, s_2 is

24   3. Finite State Automata

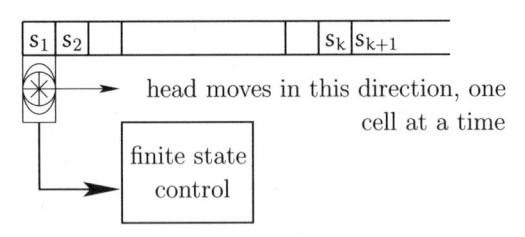

```
| s₁ | s₂ |    |    | sₖ | sₖ₊₁ |
```
head moves in this direction, one
cell at a time

finite state control

Figure 3.1   A finite automaton.

read, and the automaton changes its state to one of the possible states in $\delta(q, s_2)$. This process is iterated until the automaton reaches the first cell which is empty. Finally, after having read the whole string, the automaton is in some state, say r. If $r \in F$, then the computation has been an *accepting* one, otherwise, the string s is *rejected* (拒否される).

Now, we see what the problem is in defining the language accepted by an NFA. On input a string s, there are many possible computations. Some of these computations may finish in an accepting state and some may not. We therefore define the language accepted by an NFA as follows.

Definition 3.4   *Let $\mathcal{A} = [\Sigma, Q, \delta, q_0, F]$ be an NFA. The language $L(\mathcal{A})$ accepted by $\mathcal{A}$ is the set of all strings $s \in \Sigma^*$ such that there exists an accepting computation for s.*

Finally, finite automata may be conveniently represented by their *state diagram* (状態遷移図). The state diagram is a directed graph whose nodes are labeled by the states of the automaton. The edges are labeled by symbols from Σ. Let p and q be nodes. Then, there is a directed edge from p to q if and only if there exists an $a \in \Sigma$ such that $q = \delta(p, a)$ (deterministic case) or $q \in \delta(p, a)$ (nondeterministic case). Figure 3.2 below shows the state diagram of a finite automaton accepting the language $L = \{a^i b^j \mid i \geq 0,\ j \geq 0\}$. Note that, by convention, $a^0 b^0 = \lambda$.

The automaton displayed in Figure 3.2 has 3 states, i.e., $Q = \{1, 2, 3\}$.

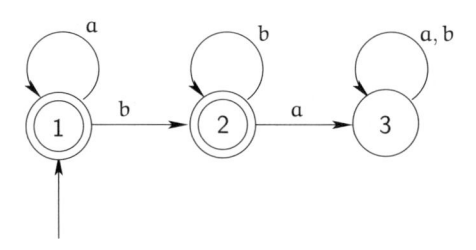

Figure 3.2 A finite automaton accepting the language $L = \{a^i b^j \mid i, j \in \mathbb{N}\}$.

The input alphabet is $\Sigma = \{a, b\}$, and the set F of final states is $\{1, 2\}$. We have marked the final states by drawing an extra circle in then. The initial state is marked by an unlabeled arrow, that is, 1 is the initial state.

3.2 Finite Automata and Regular Languages

Now that we know what finite automata are, we can answer the question that probably has already crossed the reader's mind, i.e.,

What have finite automata to do with regular languages?

We answer this question by the following theorem.

Theorem 3.1 *Let* $L \subseteq \Sigma^*$ *be any language. Then, the following three assertions are equivalent.*

(1) *There is a DFA $\mathcal{A}$ such that* $L = L(\mathcal{A})$.

(2) *There is an NFA $\mathcal{A}$ such that* $L = L(\mathcal{A})$.

(3) L *is regular.*

Proof. We show the equivalence by proving (1) implies (2), (2) implies (3), and (3) implies (1).

Claim 1. (1) *implies* (2).

This is obvious by definition, since a deterministic finite automaton is a special case of a nondeterministic one.

Claim 2. (2) *implies* (3).

Let $\mathcal{A} = [\Sigma, Q, \delta, q_0, F]$ be an NFA such that $L = L(\mathcal{A})$. We have to construct a grammar $\mathcal{G}$ generating L. Let $\mathcal{G} = [\Sigma,\ Q \cup \{\sigma\},\ \sigma,\ P]$, where P is the following set of productions:

$$P = \{\sigma \to q_0\} \cup \{p \to aq \mid a \in \Sigma,\ p, q \in Q,\ q \in \delta(p, a)\}$$
$$\cup \{p \to \lambda \mid p \in F\}.$$

Obviously, $\mathcal{G}$ is regular. We have to show $L(\mathcal{G}) = L(\mathcal{A})$. First we prove $L(\mathcal{A}) \subseteq L(\mathcal{G})$.

Let $s = a_1 \cdots a_k \in L$. Then, there exists an accepting computation of $\mathcal{A}$ for s. Let $q_0, p_1, \ldots, p_k$ be the sequence of states through which $\mathcal{A}$ goes while performing this accepting computation. Therefore, $p_1 \in \delta(q_0, a_1)$, $p_2 \in \delta(p_1, a_2)$, ..., $p_k \in \delta(p_{k-1}, a_k)$, and $p_k \in F$. Thus, $\sigma \Rightarrow q_0 \Rightarrow a_1 p_1 \Rightarrow \cdots \Rightarrow a_1 \cdots a_{k-1} p_{k-1} \Rightarrow a_1 \cdots a_k p_k \Rightarrow a_1 \cdots a_k$. Hence, we obtain $s \in L(\mathcal{G})$. The direction $L(\mathcal{G}) \subseteq L(\mathcal{A})$ can be proved analogously, and is therefore left as an exercise.

Claim 3. (3) *implies* (1).

In principle, we want to use an idea similar to the one used to prove Claim 2. But productions may have strings on their right hand side, while the transition function has to be defined over states and letters. Therefore, we first have to show a *normal form* (標準形) lemma for regular grammars.

Lemma 3.2 (Normal Form Lemma) *For every regular grammar* $\mathcal{G} = [T, N, \sigma, P]$ *there is a grammar* $\mathcal{G}'$ *such that* $L(\mathcal{G}) = L(\mathcal{G}')$ *and all productions of* $\mathcal{G}'$ *have the form* $h \to ah'$ *or* $h \to \lambda$, *where* $h, h' \in N$ *and* $a \in T$.

First, each production of the form $h \to a_1 \cdots a_k h'$, $k > 0$, is equivalently replaced by the following productions:

$$h \to a_1 h_{a_2 \cdots a_k h'},\ h_{a_2 \cdots a_k h'} \to a_2 h_{a_3 \cdots a_k h'},\ \ldots,\ h_{a_k h'} \to a_k h'.$$

Next, each production of the form $h \to a_1 \cdots a_k$, $k > 0$, is equivalently

3.2 Finite Automata and Regular Languages

replaced by the following productions:

$h \to a_1 h_{a_2 \cdots a_k}$, $h_{a_2 \cdots a_k} \to a_2 h_{a_3 \cdots a_k}$, ..., $h_{a_k} \to a_k h_\lambda$ and $h_\lambda \to \lambda$.

Finally, we have to deal with productions of the form $h \to h'$ where $h, h' \in N$. Let $\hat{\mathcal{G}} = [T, \hat{N}, \sigma, \hat{P}]$ be the grammar constructed so far. Furthermore, let $U(h) =_{df} \{h' \mid h' \in \hat{N} \text{ and } h \stackrel{*}{\Rightarrow} h'\}$. Clearly, $U(h)$ is computable. Now, we delete all productions of the form $h \to h'$ and add the following productions to $\hat{P}$. If $h' \to \lambda \in \hat{P}$ and $h' \in U(h)$, then we add $h \to \lambda$. Moreover, we add all $h \to xh''$ for all productions in $\hat{P}$ such that there is a $h' \in U(h)$ with $h' \to xh'' \in \hat{P}$.

Let $\mathcal{G}'$ be the resulting grammar. Clearly, now all productions have the desired form. One easily verifies that $L(\mathcal{G}) = L(\mathcal{G}')$. This proves the lemma.

Now, assume that we are given a grammar $\mathcal{G} = [T, N, \sigma, P]$ that is already in the normal form described in the Normal Form Lemma above. We define a DFA $\mathcal{A} = [T, Q, \delta, q_0, F]$ as follows. Let $Q = \wp(N)$ and $q_0 = \{\sigma\}$. The transition function δ is defined as

$$\delta(p, a) = \{h' \mid \exists h [h \in p \text{ and } h \to ah' \in P]\}.$$

Finally, we set

$$F = \{p \mid \exists h [h \in p \text{ and } h \to \lambda \in P]\}.$$

We leave it as an exercise to prove $L(\mathcal{A}) = L(\mathcal{G})$. ∎

Theorem 3.1 directly allows for the following corollary.

Corollary 3.1 *There is an algorithm that on input any regular grammar $\mathcal{G} = [T, N, \sigma, P]$ and any string $s \in T^*$ decides whether or not $s \in L(\mathcal{G})$.*

Proof. As the proof of Theorem 3.1 shows, given any regular grammar $\mathcal{G}$ we can construct a DFA $\mathcal{A}$ such that $L(\mathcal{G}) = L(\mathcal{A})$. Since $\mathcal{A}$ is deterministic, on input any string $s \in T^*$ it either accepts s or it rejects it. ∎

3. Finite State Automata

Next, we illustrate the transformation of a grammar into a DFA by an example. Let $\mathcal{G} = [\{a, b\}, \{\sigma, h\}, \sigma, P]$ be the grammar given, where $P = \{\sigma \to ab\sigma,\ \sigma \to h,\ h \to aah,\ h \to \lambda\}$. First, we have to transform $\mathcal{G}$ into an equivalent grammar in normal form. Thus, we obtain

$$\hat{P} = \{\sigma \to ah_{b\sigma},\ h_{b\sigma} \to b\sigma,\ \sigma \to h,\ h \to ah_{ah},$$
$$h_{ah} \to ah,\ h \to \lambda\}.$$

Now, $U(\sigma) = \{h\}$, and $U(h) = \emptyset$. Thus, we delete the production $\sigma \to h$ and replace it by $\sigma \to ah_{ah}$ and $\sigma \to \lambda$. Summarizing this construction, we now have the following set P' of productions

$$P' = \{\sigma \to ah_{b\sigma},\ h_{b\sigma} \to b\sigma,\ \sigma \to ah_{ah},\ \sigma \to \lambda,\ h \to ah_{ah},$$
$$h_{ah} \to ah,\ h \to \lambda\}$$

as well as the following set N' of nonterminals

$$N' = \{\sigma,\ h_{ah},\ h_{b\sigma},\ h\}.$$

Thus, our automaton has 16 states, i.e.,
$\{\emptyset,\ \{\sigma\},\ \{h_{ah}\},\ \{h_{b\sigma}\},\ \{h\},\ \{\sigma, h_{ah}\},\ \{\sigma, h_{b\sigma}\},\ \{\sigma, h\},\ \{h_{ah}, h_{b\sigma}\},\ \{h_{ah}, h\},$
$\{h, h_{b\sigma}\}, \{\sigma,\ h_{ah},\ h_{b\sigma}\},\ \{\sigma,\ h_{ah},\ h\},\ \{\sigma,\ h_{b\sigma},\ h\},\ \{h_{ah},\ h_{b\sigma},\ h\},$
$\{\sigma,\ h_{ah},\ h_{b\sigma},\ h\}\}$.

The set of final states is

$$F = \{\{\sigma\},\ \{h\}\}$$
$$\cup \{\{\sigma, h_{ah}\},\ \{\sigma, h_{b\sigma}\},\ \{\sigma, h\},\ \{h_{ah}, h\},\ \{h, h_{b\sigma}\},\ \{\sigma,\ h_{ah},\ h_{b\sigma}\},$$
$$\{\sigma,\ h_{ah},\ h\}, \{\sigma,\ h_{b\sigma},\ h\},\ \{h_{ah},\ h_{b\sigma},\ h\},\ \{\sigma,\ h_{ah},\ h_{b\sigma},\ h\}\}.$$

Finally, we have to compute δ. We illustrate this part here only for two states, the rest is left as an exercise. For computing $\delta(\{\sigma\}, a)$, we have to consider the set of all productions having σ on the left hand side and a

3.2 Finite Automata and Regular Languages

on the right hand side. There are two such productions, i.e., $\sigma \to ah_{b\sigma}$ and $\sigma \to ah_{ah}$; thus $\delta(\{\sigma\}, a) = \{h_{b\sigma}, h_{ah}\}$. Since there is no production having σ on the left hand side and b on the right hand side, we obtain $\delta(\{\sigma\}, b) = \emptyset$. Analogously, we get $\delta(\{h_{b\sigma}, h_{ah}\}, a) = \{h\}$, and $\delta(\{h_{b\sigma}, h_{ah}\}, b) = \{\sigma\}$, $\delta(\{h\}, a) = \{h_{ah}\}$, $\delta(\{h\}, b) = \emptyset$, $\delta(\{h_{ah}\}, a) = \{h\}$, $\delta(\{h_{ah}\}, b) = \emptyset$ as well as $\delta(\emptyset, a) = \delta(\emptyset, b) = \emptyset$.

Looking at the transitions computed so far, we see that none of the other states appeared, and thus they can be ignored.

Exercise 3.1 *Complete the calculation of the automaton above and draw its state diagram.*

Finally, we show how to use finite automata to prove that particular languages are *not regular*. Consider the following grammar

$$\mathcal{G} = [\{a, b\}, \{\sigma\}, \sigma, \{\sigma \to a\sigma b, \sigma \to \lambda\}]$$

Then, $L(\mathcal{G}) = \{a^n b^n \mid n \in \mathbb{N}\}$. Clearly, $\mathcal{G}$ is not regular, but this does not prove that there is no regular grammar $\mathcal{G}'$ at all such that $L(\mathcal{G}) = L(\mathcal{G}')$.

Theorem 3.2 *The language* $L = \{a^n b^n \mid n \in \mathbb{N}\}$ *is not regular.*

Proof. Suppose the converse. Then, by Theorem 3.1 there must be a DFA $\mathcal{A} = [\Sigma, Q, \delta, q_0, F]$ such that $L(\mathcal{A}) = L$. Clearly, $\{a, b\} \subseteq \Sigma$, and thus $\delta(q_0, a^i)$ must be defined for all $i \in \mathbb{N}$. However, there are only finitely many states, but infinitely many i, and hence there must exist i, j such that $i \neq j$ but $\delta(q_0, a^i) = \delta(q_0, a^j)$. Since $\mathcal{A}$ is deterministic, $\delta(q_0, a^i) = \delta(q_0, a^j)$ implies $\delta(q_0, a^i b^i) = \delta(q_0, a^j b^i)$. Let $q = \delta(q_0, a^i b^i)$; then, if $q \in F$ we directly obtain $a^j b^i \in L$, a contradiction, since $i \neq j$. But if $q \notin F$, then $a^i b^i$ is rejected. This is again a contradiction, since $a^i b^i \in L$. Therefore, the language L is not regular. ∎

Last but not least, we need some more advanced problems to see if we have understood so far everything correctly, and here they come.

Exercise 3.2 Let $\Sigma = \{0, 1\}$, and let the formal language L_1 be defined as $L_1 = \{w \in \Sigma^+ \mid |w| \text{ is divisible by } 4\}$. Prove or disprove L_1 to be regular.

Exercise 3.3 Let L_k be the language defined in Problem 2.3. Show that every DFA $\mathcal{A} = [\{a, b\}, Q, \delta, q_0, F]$ with $L(\mathcal{A}) = L_k$ satisfies card$(Q) > k$.

3.3 Problem Set 3

Problem 3.1 Let $w \in \{0, 1\}^*$ be any given string. We define

$$L_w = \{uwv \mid u, v \in \{0, 1\}^*\},$$

i.e., L_w is the set of all strings over $\{0, 1\}$ containing w as a *substring*.

(1) Construct an NFA $\mathcal{A} = [\{0, 1\}, Q, \delta, q_0, F]$ accepting L_{110011}.

(2) Transform $\mathcal{A}$ into a DFA that also accepts L_{110011}.

Problem 3.2 Consider the following NFA $\mathcal{A} = [\{0, 1\}, Q, \delta, q_0, F]$, where $Q = \{q_0, q_1, q_2, q_3, q_4, q_5, q_6, q_7\}$ and $F = \{q_7\}$, and

$\delta(q_0, 0) = \{q_0, q_1\}, \quad \delta(q_0, 1) = \{q_0, q_2\}, \quad \delta(q_1, 0) = \{q_3\},$

$\delta(q_2, 1) = \{q_4\}, \quad \delta(q_3, 0) = \{q_5\}, \quad \delta(q_4, 0) = \{q_6\},$

$\delta(q_5, 1) = \{q_7\}, \quad \delta(q_6, 0) = \{q_7\}$.

(1) Construct a DFA $\mathcal{A}'$ accepting the same language as $\mathcal{A}$.

(2) Draw the state diagrams of $\mathcal{A}$ and $\mathcal{A}'$, respectively.

(3) Construct a regular grammar $\mathcal{G}$ such that $L(\mathcal{G}) = L(\mathcal{A})$.

Problem 3.3 Let $\Sigma = \{a, b\}$, and consider $L_{pal} = \{w \in \Sigma^* \mid w = w^T\}$. Prove or disprove L_{pal} to be regular.

4 Characterizations of $\mathcal{REG}$

We start this chapter by proving an algebraic characterization for $\mathcal{REG}$.

4.1 Nerode's Theorem

Recall that a binary relation over a set $X \neq \emptyset$ is an equivalence relation if it is reflexive, symmetric and transitive (cf. Definition 1.2). As shown in Problem 1.2, the set of all equivalence classes forms a partition of the underlying set X. Additionally, we need the following definitions.

Definition 4.1 *Let $X \neq \emptyset$ be any set and $\rho \subseteq X \times X$ be any equivalence relation. The relation ρ is said to be of* finite rank (有限の階数) *if X/ρ is finite.*

Next, we define the *Nerode relation* (ネロード関係) (cf. Nerode [26]). Let Σ be any alphabet, let $L \subseteq \Sigma^*$ be any language and $v, w \in \Sigma^*$. We set

$$v \sim_L w \quad \text{iff} \quad \forall u \in \Sigma^*[vu \in L \iff wu \in L] . \tag{4.1}$$

Exercise 4.1 *Show that $\sim_L$ is an equivalence relation.*

Let $\sim$ be any equivalence relation over Σ^*. We call $\sim$ *right invariant*

$$(右不変) \quad \text{if} \quad \forall u, v, w \in \Sigma^*[u \sim v \quad \text{implies} \quad uw \sim vw] . \tag{4.2}$$

Exercise 4.2 *Show $\sim_L$ to be right invariant.*

Now, we are ready to prove our first characterization in this chapter.

Theorem 4.1 (Nerode's Theorem [26]) Let Σ be any alphabet and $L \subseteq \Sigma^*$ be any language. Then $L \in \mathcal{REG}$ if and only if $\sim_L$ is of finite rank.

Proof. Necessity. Let $L \in \mathcal{REG}$. Then there is a DFA $\mathcal{A} = [\Sigma, Q, \delta, q_0, F]$ such that $L(\mathcal{A}) = L$. We define the following relation $\sim$ over Σ^*:

$$v \sim w \quad \text{if} \quad \delta(q_0, v) = \delta(q_0, w) \quad \text{for all } v, w \in \Sigma^* . \tag{4.3}$$

Claim 1. The relation $\sim$ is an equivalence relation.

Since $=$ is an equivalence relation, we can directly conclude that $\sim$ is an equivalence relation, too.

Claim 2. The relation $\sim$ is of finite rank.

Since $\text{card}(\Sigma^*/\sim) \leq \text{card}(Q)$ and $\text{card}(Q)$ is finite (cf. Definition 3.2), the relation $\sim$ must be of finite rank.

Claim 3. $v \sim w$ implies $v \sim_L w$ for all $v, w \in \Sigma^*$.

Let $u, v, w \in \Sigma^*$ such that $v \sim w$. We must show: $vu \in L \iff wu \in L$. Since $L = L(\mathcal{A})$, it suffices to prove that $vu \in L(\mathcal{A}) \iff wu \in L(\mathcal{A})$. By Definition 3.3 we know that $vu \in L(\mathcal{A}) \iff \delta(q_0, vu) \in F$. Because of $v \sim w$ we also have $\delta(q_0, v) = \delta(q_0, w)$. Thus, by Lemma 3.1 we have

$$\delta(q_0, vu) = \delta(\delta(q_0, v), u) = \delta(\delta(q_0, w), u) = \delta(q_0, wu) .$$

Hence, $\delta(q_0, vu) \in F \iff \delta(q_0, wu) \in F$, and thus $vu \in L \iff wu \in L$.

Claim 1 through 3 directly imply that $\sim_L$ is of finite rank.

Sufficiency. Let $L \subseteq \Sigma^*$ be such that $\sim_L$ is of finite rank. We have to show that $L \in \mathcal{REG}$. By Theorem 3.1 it suffices to construct a DFA $\mathcal{A} = [\Sigma, Q, \delta, q_0, F]$ such that $L = L(\mathcal{A})$. We set

- $Q = \Sigma^*/_{\sim_L}$,
- $q_0 = [\lambda]$,
- $\delta([w], x) = [wx]$ for all $[w] \in Q$ and all $x \in \Sigma$,
- $F = \{[w] \mid w \in L\}$.

Exercise 4.3 *Show that the definition of δ does not depend on the representative (代表元) of the equivalence class* [w].

Claim 4. $\delta([\lambda], w) = [w]$ *for all* $w \in \Sigma^*$.

The claim is proved inductively. For the induction basis we have

$$\delta([\lambda], x) = [\lambda x] = [x] , \quad \text{since } \lambda x = x \text{ for all } x \in \Sigma .$$

Now, suppose as induction hypothesis $\delta([\lambda], w) = [w]$ and let $x \in \Sigma$. We have to show $\delta([\lambda], wx) = [wx]$. So, we calculate

$$\delta([\lambda], wx) = \delta(\delta([\lambda], w), x) = \delta([w], x) = [wx] ,$$

where the first equality is by the extension of the definition of δ to strings (cf. Lemma 3.1), the second one by the induction hypothesis, and the last one by the definition of δ. This proves Claim 4.

By construction and Claim 4, we directly obtain

$$w \in L(\mathcal{A}) \iff \delta([\lambda], w) \in F$$
$$\iff [w] \in F$$
$$\iff w \in L .$$

This shows the sufficiency and thus the theorem is shown. ∎

4.2 Regular Expressions

Next, we construct a language such that each string of it can be regarded as a generator of a regular language. Let Σ be any fixed alphabet. We set

$$\underline{\Sigma} = \{\underline{x} \mid x \in \Sigma\} \cup \{\oslash, \wedge, \vee, \cdot, \langle, \rangle, (,)\} , \tag{4.4}$$

that is, for every $x \in \Sigma$ we introduce a *new* symbol called $\underline{x}$ and, additionally, we introduce the symbols $\oslash, \wedge, \vee, \cdot, \langle, \rangle, (,)$. Note that the comma is a *meta symbol* (メタ記号).

4. Characterizations of $\mathcal{REG}$

Next we set $\mathcal{G}_{reg} = [\underline{\Sigma}, \{\sigma\}, \sigma, P]$, where P is defined as follows:

$P = \{\sigma \to \underline{x} \mid x \in \Sigma\}$
$\cup \{\sigma \to \oslash,\ \sigma \to \Lambda,\ \sigma \to (\sigma \vee \sigma),\ \sigma \to (\sigma \cdot \sigma),\ \sigma \to \langle \sigma \rangle\}$.

We call $L(\mathcal{G}_{reg})$ the language of *regular expressions* (正規表現) over Σ.

So far, we have defined the *syntax* (構文) of regular expressions. Next, we define their *semantics* (意味). Let $T, T_1, T_2 \in L(\mathcal{G}_{reg})$, we inductively define $L(T)$ as follows.

Induction Basis: $L(\underline{x}) = \{x\}$ for all $x \in \Sigma$, $L(\oslash) = \emptyset$ and $L(\Lambda) = \{\lambda\}$.

Induction Step: $L(T_1 \vee T_2) = L(T_1) \cup L(T_2)$

$L(T_1 \cdot T_2) = L(T_1)L(T_2)$

$L(\langle T \rangle) = L(T)^*$.

Theorem 4.2 *Let Σ be any fixed alphabet and let $\mathcal{G}_{reg}$ be defined as above. Then we have: A language $L \subseteq \Sigma^*$ is regular if and only if there exists a regular expression T such that $L = L(T)$.*

Proof. Sufficiency. Let $T \in L(\mathcal{G}_{reg})$ be any regular expression. We have to show that $L(T)$ is regular. This is done inductively over T.

The induction basis is obvious, since all singleton (単一要素の) languages, the empty language $\emptyset$, and the language $\{\lambda\}$ are regular.

The induction step is also clear, since by Theorem 2.1 we already know that the regular languages are closed under union, product and Kleene closure. This proves the sufficiency.

Next, we define what is meant by prefix, suffix and substring, respectively. Let $w, y \in \Sigma^*$. We write $w \sqsubseteq y$ if there exists a string $v \in \Sigma^*$ such that $wv = y$. If $w \sqsubseteq y$, then we call w a *prefix* (接頭辞) of y. If $v \neq \lambda$, then w is said to be a *proper prefix* (真の接頭辞) of y. In this case we write $w \sqsubset y$. Analogously, we call w a *suffix* (接尾辞) of y if there exists a string $v \in \Sigma^*$ such that $vw = y$ and if $v \neq \lambda$, then w is said to be a *proper suffix* (真の接尾辞) of y. Finally, w is said to be a *substring* (部分文字列) of y if there exist strings $u, v \in \Sigma^*$ such that $uwv = y$.

For showing the necessity, let $L \in \mathcal{REG}$ be arbitrarily fixed. We have to construct a regular expression T such that $L = L(T)$.

Since $L \in \mathcal{REG}$, there is a DFA $\mathcal{A} = [\Sigma, Q, \delta, q_1, F]$ such that $L = L(\mathcal{A})$. Let $Q = \{q_1, \ldots, q_m\}$ and let $F \subseteq Q$. We distinguish the following cases.

Case 1. $F = \emptyset$.

In this case we can directly conclude $L = L(\mathcal{A}) = \emptyset$. Thus, clearly there exists $T \in L(\mathcal{G}_{reg})$ such that $L = L(T) = \emptyset$, i.e., $T = \oslash$ does it.

Case 2. $F \neq \emptyset$.

By Definition 3.3 we can write

$$L(\mathcal{A}) = \bigcup_{q \in F} \{s \mid s \in \Sigma^* \text{ and } \delta(q_1, s) = q\} .$$

Now, we can decompose $\mathcal{A}$ into DFAs $\mathcal{A}_q = [\Sigma, Q, \delta, q_1, \{q\}]$, where $q \in F$ and F is the set of accepting states from the DFA $\mathcal{A}$. Thus, we obtain $L(\mathcal{A}) = \bigcup_{q \in F} L(\mathcal{A}_q)$. Hence, it suffices to construct for each DFA $\mathcal{A}_q$ a regular expression T_q such that $L(T_q) = L(\mathcal{A}_q)$, since then $L(\mathcal{A}) = L\left(\bigvee_{q \in F} T_q\right)$.

Let $i, j, k \leq m$ and define

$$L_{i,j}^k = \{s \mid s \in \Sigma^* \text{ and } \delta(q_i, s) = q_j \text{ and}$$
$$\forall t \forall r [\lambda \neq r \sqsubset s \text{ and } \delta(q_i, r) = q_t \text{ implies } t \leq k]\} .$$

Next, we show that for each $L_{i,j}^k$ there is a regular expression $T_{i,j}^k$ such that $L_{i,j}^k = L(T_{i,j}^k)$. This will complete the proof, since $L(\mathcal{A}_q) = L_{1,n}^m$, provided $q = q_n$, $n \leq m$.

The proof is by induction on k. For the induction basis, let $k = 0$. Then

$$L_{i,j}^0 = \{s \mid s \in \Sigma^* \text{ and } \delta(q_i, s) = q_j \text{ and }$$
$$\forall t \forall r [\lambda \neq r \sqsubset s \text{ and } \delta(q_i, r) = q_t \text{ implies } t \leq 0]\}$$
$$= \{x \mid x \in \Sigma \text{ and } \delta(q_i, x) = q_j\}.$$

That means, either we have $L_{i,j}^0 = \emptyset$ or $L_{i,j}^0$ is a finite set of strings of length 1, i.e., $\text{card}(L_{i,j}^0) \leq \text{card}(\Sigma)$. If $L_{i,j}^0 = \emptyset$ we set $T_{i,j}^0 = \oslash$ and we are done. If $L_{i,j}^0 \neq \emptyset$, say $L_{i,j}^0 = \{x^{(1)}, \ldots, x^{(z)}\}$, where $z \leq \text{card}(\Sigma)$, we set

$$T_{i,j}^0 = \underline{x}^{(1)} \vee \cdots \vee \underline{x}^{(z)},$$

and, by the inductive definition of L(T), we directly get $L_{i,j}^0 = L(T_{i,j}^0)$. This proves the induction basis.

Now, assume the induction hypothesis that $L_{i,j}^k$ is regular and there is a regular expression $T_{i,j}^k$ such that $L_{i,j}^k = L(T_{i,j}^k)$ for all $i, j \leq m$ and $k < m$.

For the induction step, it suffices to construct a regular expression $T_{i,j}^{k+1}$ such that $L_{i,j}^{k+1} = L(T_{i,j}^{k+1})$.

This is done via the following lemma.

Lemma 4.1 $L_{i,j}^{k+1} = L_{i,j}^k \cup L_{i,k+1}^k \left(L_{k+1,k+1}^k\right)^* L_{k+1,j}^k$

The "$\supseteq$" direction is obvious. For showing

$$L_{i,j}^{k+1} \subseteq L_{i,j}^k \cup L_{i,k+1}^k \left(L_{k+1,k+1}^k\right)^* L_{k+1,j}^k$$

let $s \in L_{i,j}^{k+1}$, say $s = x_1 x_2 \cdots x_\ell$. We consider the sequence of states reached when successively processing s, i.e., $q_i, q^{(1)}, q^{(2)}, \ldots, q^{(\ell-1)}, q_j$. We distinguish the following cases.

Case α. $q_i, q^{(1)}, q^{(2)}, \ldots, q^{(\ell-1)}, q_j$ *does not contain* q_{k+1}.

Then, clearly $s \in L_{i,j}^k$ and we are done.

Case β. $q_i, q^{(1)}, q^{(2)}, \ldots, q^{(\ell-1)}, q_j$ *contains* q_{k+1}.

Now, we may depict the situation as follows:

$$q_i \xrightarrow{u} q_{k+1} \xrightarrow{s_1} q_{k+1} \xrightarrow{s_2} q_{k+1} \cdots q_{k+1} \xrightarrow{s_\mu} q_{k+1} \xrightarrow{v} q_j.$$

More formally, let u be the shortest non-empty prefix of s such that $\delta(q_i, u) = q_{k+1}$ and let $s_1, \ldots, s_\mu$ and v be all strings such that

$$s = u s_1 s_2 \cdots s_\mu v \quad \text{and}$$

$$\delta(q_i, u) = \delta(q_i, u s_1) = \delta(q_i, u s_1 s_2) = \cdots = \delta(q_i, u s_1 s_2 \cdots s_\mu) = q_{k+1}.$$

Hence, $u \in L_{i,k+1}^k$ and $s_1, \ldots, s_\mu \in L_{k+1,k+1}^k$ as well as $v \in L_{k+1,j}^k$. Consequently, we arrive at

$$s \in L_{i,j}^k \cup L_{i,k+1}^k \left(L_{k+1,k+1}^k \right)^* L_{k+1,j}^k .$$

Therefore, we set

$$T_{i,j}^{k+1} = T_{i,j}^k \vee T_{i,k+1}^k \cdot \langle T_{k+1,k+1}^k \rangle \cdot T_{k+1,j}^k ,$$

and the induction step is shown. So, Theorem 4.2 is proved. ∎

We succeeded to characterize the regular languages by regular expressions. Besides their mathematical beauty, regular expressions are also of fundamental practical importance. We shall come back to this point in the next chapter.

4.3 The Pumping Lemma for Regular Languages

We continue with another important property of regular languages which is often stated as *Pumping Lemma* (反復補題) or sru-Theorem.

Lemma 4.2 (Pumping Lemma for $\mathcal{REG}$) *For every regular language* L *there is a number* $k \in \mathbb{N}$ *such that for all* $w \in L$ *with* $|w| \geq k$ *there are strings* s, r, u *such that* $w = sru$, $r \neq \lambda$ *and* $sr^i u \in L$ *for all* $i \in \mathbb{N}$.

Proof. If L is finite then we can choose k such that $|w| < k$ for all $w \in L$. Thus, $\{w \mid w \in L \text{ and } |w| \geq k\} = \emptyset$, and the assertion of the lemma trivially holds.

Next, let L be any infinite regular language. Then there exists a DFA $\mathcal{A} = [\Sigma, Q, \delta, q_0, F]$ such that $L = L(\mathcal{A})$. Let $n = \text{card}(Q)$. We show that $k = n + 1$ satisfies the conditions of the lemma.

Let $w \in L$ such that $|w| \geq k$. Then there must be strings $s, r, u \in \Sigma^*$ such that $r \neq \lambda$, $w = sru$ and $q_* =_{df} \delta(q_0, s) = \delta(q_0, sr)$. Consequently, $q_* = \delta(q_0, sr^i)$ for all $i \in \mathbb{N}$. Because of $\delta(q_0, sru) \in F$, we conclude $\delta(q_0, sr^i u) \in F$, and thus the Pumping Lemma is shown. ∎

Exercise 4.4 *Prove the Pumping Lemma directly by using regular grammars instead of finite automata.*

As an example for the application of the pumping lemma, we consider again the language $L = \{a^i b^i \mid n \in \mathbb{N}^+\}$. We claim that $L \notin \mathcal{REG}$. Suppose the converse. Then, by the Pumping Lemma, there must be a number k such that for all $w \in L$ with $|w| \geq k$ there are strings s, r, u such that $w = sru$, $r \neq \lambda$ and $sr^i u \in L$ for all $i \in \mathbb{N}$. So, let $w = a^k b^k = sru$. We distinguish the following cases.

Case 1. $r = a^i b^j$ for some $i, j \in \mathbb{N}^+$.

Then we get $srru = a^{k-i} a^i b^j a^i b^j b^{k-j}$, i.e., $srru \notin L$, a contradiction.

Case 2. $r = a^i$ for some $i \in \mathbb{N}^+$.

Then, we directly get $srru = a^{k-i} a^i a^i b^k = a^{k+i} b^k$. Again, $srru \notin L$, a contradiction.

Case 3. $r = b^i$ for some $i \in \mathbb{N}^+$.

This case can be handled analogously to Case 2. Thus, $L \notin \mathcal{REG}$.

Note that the pumping lemma provides a necessary condition for a language to be regular. The condition is *not sufficient*.

Exercise 4.5 *Show that there is a language $L \notin \mathcal{REG}$ such that L satisfies the conditions of the Pumping Lemma.*

Additionally, using the ideas developed so far, we can show another important property.

Theorem 4.3 *There is an algorithm which on input any regular grammar $\mathcal{G}$ decides whether or not $L(\mathcal{G})$ is infinite or finite.*

Proof. Let $\mathcal{G}$ be a regular grammar. The algorithm first constructs a deterministic finite automaton $\mathcal{A} = [\Sigma, Q, \delta, q_0, F]$ such that $L(\mathcal{G}) = L(\mathcal{A})$. Let $\text{card}(Q) = n$.

Then, the algorithm checks whether or not there is a string s such that $n + 1 \leq |s| \leq 2n + 2$ with $s \in L(\mathcal{A})$.

If not, then output "$L(\mathcal{G})$ is finite."

Otherwise output "$L(\mathcal{G})$ is infinite."

It remains to show that the algorithm always terminates and works correctly.

Since the construction of a deterministic finite automaton from a given grammar is constructive, this step will always terminate. Furthermore, the number of strings s satisfying $n + 1 \leq |s| \leq 2n + 2$ is finite. Thus, the test will terminate, too.

If there is such a string s with $n + 1 \leq |s| \leq 2n + 2$ and $s \in L(\mathcal{A})$, then by the proof of the Pumping Lemma we can directly conclude that $L(\mathcal{G})$ is infinite.

Finally, suppose there is no such string s but $L(\mathcal{G})$ is infinite. Then, there must be at least one string $w \in L(\mathcal{G})$ with $|w| > 2n+2$. Since $\text{card}(Q) = n$, it is obvious that $\mathcal{A}$, when processing w must reach some states more than once. Now, we can cut off sufficiently many substrings of w that transfer one of the states into itself. The resulting string w' must then have a length between $n+1$ and $2n+2$, a contradiction. So, we can conclude that $w \in L(\mathcal{G})$ implies $|w| \leq n$, and thus $L(\mathcal{G})$ is finite. ∎

Now, apply the knowledge gained so far to solve the following exercises.

Exercise 4.6 *Prove or disprove: there is an algorithm which on input any regular grammar $\mathcal{G}$ decides whether or not $L(\mathcal{G}) = \emptyset$.*

Exercise 4.7 *Use Theorem 4.1 to show that $L = \{a^n b^n \mid n \in \mathbb{N}\} \notin \mathcal{REG}$.*

Exercise 4.8 *Prove or disprove: there is an algorithm which on input any regular grammars $\mathcal{G}_1$, $\mathcal{G}_2$ decides whether or not $L(\mathcal{G}_1) \cap L(\mathcal{G}_2) = \emptyset$.*

Exercise 4.9 *Prove or disprove: there is an algorithm which on input any regular grammars $\mathcal{G}_1$, $\mathcal{G}_2$ decides whether or not $L(\mathcal{G}_1) \subseteq L(\mathcal{G}_2)$.*

Exercise 4.10 *Prove or disprove: there is an algorithm which on input any regular grammars $\mathcal{G}_1$, $\mathcal{G}_2$ decides whether or not $L(\mathcal{G}_1) = L(\mathcal{G}_2)$.*

4.4 Problem Set 4

Problem 4.1 Prove the following extension of Nerode's theorem. Let Σ be any alphabet and $L \subseteq \Sigma^*$ be any language. Then $L \in \mathcal{REG}$ if and only if there is a right invariant equivalence relation $\approx$ such that $\approx$ is of finite rank and L is the union of some equivalence classes with respect to $\approx$.

Problem 4.2 Prove or disprove $\mathcal{REG}$ to be closed under

(1) *transposition*, i.e., for all $L \in \mathcal{REG}$ we have $L^\mathsf{T} \in \mathcal{REG}$.

(2) *set difference*, i.e., for all $L_1, L_2 \in \mathcal{REG}$ we have $L_1 \setminus L_2 \in \mathcal{REG}$.

Problem 4.3 Prove or disprove the language of regular expressions to be regular.

5
Regular Expressions in UNIX

In this chapter we describe applications of the theory developed so far. In particular, we shall have a look at regular expressions as used in UNIX. Before we see the applications, we introduce the UNIX notation for extended regular expressions. Note that the full UNIX extensions allow to express certain non-regular languages. We do not consider these extensions here. Also note that the "basic" UNIX syntax for regular expressions is now defined as obsolete by POSIX, but is still widely used for the purposes of backwards compatibility. Most UNIX utilities (`grep, sed` ...) use it by default. Here, `grep` stands for "**g**lobal search for a **r**egular **e**xpression and **p**rint out matched lines," and `sed` for "**s**tream **ed**itor."

Most real applications deal with the ASCII[†] character set that contains 128 characters. Suppose we have the alphabet $\{a, b\}$ and want to express "any character." Then we could simply write $\underline{a} \lor \underline{b}$. However, if we have 128 characters, expressing "any character" in the same way results in a very long expression, since we have to list all characters. Thus, UNIX regular expressions allow us to write *character classes* (文字クラス) to represent large sets of characters succinctly. The rules for character classes are:

- The symbol . (dot) stands for any single character.

[†] ASCII stands for "American Standard Code for Information Interchange." It has been introduced in 1963, became a standard in 1967, and was last updated in 1986. It uses a 7-bit code.

5. Regular Expressions in UNIX

- The expression $[a_1 a_2 \cdots a_k]$ stands for the regular expression

$$a_1 \vee a_2 \vee \cdots \vee a_k .$$

  So, it saves half the characters we have to write (we omit the $\vee$).

- Between the square braces we can put a range of the form $a-d$ to mean all the characters from a to d in the ASCII sequence. Thus, $[a-z]$ matches any lowercase letter. So, if we want to express the set of all letters and digits, we can shortly write $[A-Za-z0-9]$.

- The braces [,] or other characters that have a special meaning in UNIX regular expressions are represented by preceding them with a backslash \.

- [^] matches a single character that is not contained within the brackets. For example, $[\char`\^ a-z]$ matches any single character that is not a lowercase letter.

- ^matches the start of the line and $ the end of the line, respectively.

- \(\) is used to treat the expression enclosed within the brackets as a single block.

- * matches the last block zero or more times, i.e., it stands for $\langle \ \rangle$ in our notation. For example, we can write \(abc \)* to match the empty string λ, or abc, abcabc, abcabcabc, and so on.

- \{x, y\} matches the last block at least x and at most y times. Consequently, a\{3, 5\}matches aaa, aaaa or aaaaa.

- There is no representation of the set union operator in this syntax.

The more modern UNIX regular expressions can often be used with modern UNIX utilities by including the command line flag "-E".

POSIX' extended regular expressions are similar in syntax to the traditional UNIX regular expressions, with some exceptions. The following meta-characters are added:

- | is used instead of the operator ∨ to denote union.
- The operator ? means "zero or one of" the last block, thus b$\underline{a}$? matches b or b$\underline{a}$.
- The operator + means "one or more of." Thus, R+ is a shorthand for R⟨R⟩ in our notation.
- Interestingly, backslashes are removed in the more modern UNIX regular expressions, i.e., \(\) becomes ( ) and \{ \} becomes { }.
- Note that we can also omit the second argument in {x,y} if x = y, thus $\underline{a}$\{5\} stands for $\underline{a}\underline{a}\underline{a}\underline{a}\underline{a}$.

Also, one just uses the usual characters to write the regular expressions down and not the underlined symbols we had used, i.e., one simply writes a instead of $\underline{a}$. Furthermore, the · is also omitted.

Note that Perl's regular expressions have a much richer syntax than even the extended POSIX regular expressions. This syntax has also been used in other utilities and applications. Although still named "regular expressions", the Perl extensions give an expressive power that far exceeds the regular languages. Next, we look at some applications.

5.1 Lexical Analysis

One of the oldest applications of regular expressions was in specifying the component of a compiler called *lexical analyzer* (字句解析器). This component scans the source program and recognizes *tokens* (トークン), i.e., those substrings of consecutive characters that belong together logically. Keywords and identifiers are common examples of tokens but there are many others.

The UNIX command `lex` and its GNU version `flex`, accept as input a list of regular expressions, in the UNIX style, each followed by a bracketed section of code indicating what the lexical analyzer should do when it finds an instance of that token. Such a facility is called *lexical-analyzer generator* (字句解析器の生成器), since it takes as input a high-level description of a lexical analyzer and produces from it a function working as lexical analyzer.

Commands like `lex` and `flex` are very useful, since the regular expression notation is exactly as powerful as needed to describe tokens. These commands are able to use the regular-expression-to-DFA algorithm to generate an efficient function that breaks source programs into tokens. The main advantage is the code writing, since regular expressions are much easier to write than a DFA. Also, if we need to change something, then changing a regular expression is often quite simple, while changing the code implementing a DFA can be a nightmare.

Example 5.1 Consider the following regular expression:

$(0|1)^*1(0|1)(0|1)(0|1)(0|1)(0|1)(0|1)(0|1)(0|1)(0|1)(0|1)(0|1)(0|1)(0|1)(0|1)$.

The reader may try to design a deterministic finite automaton that accepts the language described by this regular expression. But please plan to use the whole week-end for doing it, since it may have much more states than you may expect. For seeing this, we should start with much shorter versions of this regular expression, i.e., by looking at

$(0|1)^*1(0|1)$

$(0|1)^*1(0|1)(0|1)$

$(0|1)^*1(0|1)(0|1)(0|1)$

and so on. Now, try it yourself. Provide a regular expression such that its deterministic finite automaton has roughly 32,000,000 many states.

Next, let us come back to lexical analyzers. Figure 5.1 provides an example of partial input to the `lex` command.

```
else                    {return(ELSE);}
[A − Za − z][A − Za − z0 − 9]*    {code to enter the found identifier
                        {in the symbol table;
                        {return(ID);
                        }
>=                      {return(GE);}
=                       {return(EQ);}
...
```

Figure 5.1 A sample of lex input.

The first line handles the keyword else and the action is to return a symbolic constant, i.e., ELSE in this example.

Line 2 contains a regular expression describing identifiers: a letter followed by zero or more letters and/or digits. The action is to enter the found identifier to the symbol table if not already there and to return the symbolic constant ID, which has been chosen here to represent identifiers.

The third line is for the sign >=, a two character operator. The last line is for the sign =, a one character operator. In practice, there would appear expressions describing each of the keywords, each of the signs and punctuation symbols like commas and parentheses, and families of constants such as numbers and strings[†].

5.2 Finding Patterns in Text

We often use a text editor (or the UNIX program grep) to find some text in a file (e.g., the place where we defined regular expressions). A closely related problem is to filter or to find suitable web-pages.

How do these tools work?

Commonly two algorithms are used: Knuth-Morris-Pratt (KMP) and

---

[†] The example in Figure 5.1 is from [15]. More examples are provided in the man-page of flex

Boyer-Moore (BM). Both use similar ideas and take linear time: $O(m+n)$, where m is the length of the search string, and n is the length of the file. Both only test whether certain characters are equal or unequal, they do not do any complicated arithmetic on characters.

BM is a little faster in practice, but more complicated. KMP is simpler, so we shortly discuss it here. Suppose we want to grep the string saso. The idea is to construct a DFA which stores in its current state the information we need about the string seen so far. Suppose the string seen so far is "nauvwsa," then we need to know two things.

1. Did we already match the string we are looking for (saso)?
2. If not, could we possibly be in the middle of a match?

If we are in the middle of a match, we also need to know how much of the string we are looking for we have already seen. Thus, our states should be partial matches to the pattern. The possible partial matches to saso are λ, s, sa, sas, or the complete match saso itself. That is, we have to consider all prefixes including λ of our string. If the string has length m, then there are $m + 1$ such prefixes. Thus, we need $m + 1$ states for our DFA to memorize the possible partial matches. The start and the accept state are obvious, i.e., the empty match and the full match, respectively.

In general, the transition from state plus character to state is the longest string that is simultaneously a prefix of the original pattern and a suffix of the state plus character we have just seen. For example, if we have already seen sas but the next character is a, then we only have the partial match sa. Using the ideas outlined above, we are able to produce the deterministic finite automaton we are looking for. It is displayed in Figure 5.2 below.

The description given above is sufficient to get a string matching algorithm that takes time $O(m^3 + n)$. Here we need time $O(m^3)$ to build the the state table described above, and $O(n)$ to simulate it on the input file.

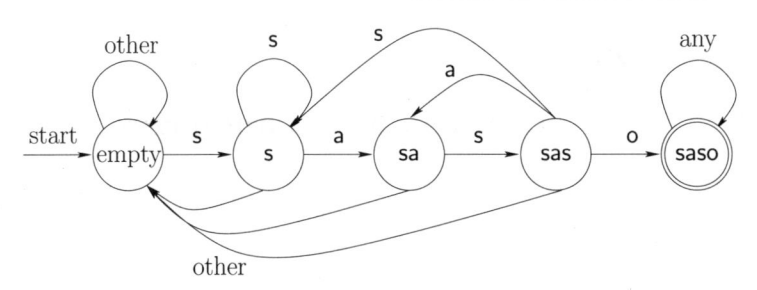

Figure 5.2 A finite automaton accepting all strings containing saso as substring.

There are two tricky points to the KMP algorithm. First, it uses an alternate representation of the state table which takes only $O(m)$ space (the one above could take $O(m^2)$). Second, it uses a complicated loop to build the whole thing in $O(m)$ time. We should have seen these tricks in earlier courses. If not, please take a look into the reference [6].

5.3 Problem Set 5

Problem 5.1 Prove or disprove the following assertions to be true:
(1) The language generated by Λ^* is empty.
(2) $abaa \in L((a|b)^*(a|bb))$.
(3) $L((a|b)^*) = L((a|b)^*|(a|b)^*)$.

Problem 5.2 Provide a regular expression generating L, where
(1) $L = \{w \mid w \in \{a,b\}^*, \text{ 3 divides } |w|\}$.
(2) The language L is the set of all strings over the alphabet $\{a,b\}$ that contain exactly two a's (e.g., $aa, abbbba \in L$ but $babbbaba \notin L$).
(3) The language L is the set of all strings over the alphabet $\{a,b\}$ that do not end with ab.

Problem 5.3 Construct a DFA accepting the language generated by the regular expression $(a|b)^*b(a|b)(a|b)$.

6
Context-Free Languages

Context-free languages were originally conceived by Noam Chomsky as a way to describe natural languages. That promise has not been fulfilled. However, context-free languages have found numerous applications in computer science, e.g., for designing programming languages, for constructing parsers for them, and for mark-up languages. The reason for the wide use of context-free grammars is that they represent the best compromise between power of expression and ease of implementation. Regular languages are too weak for many applications, since they cannot describe situations such as checking that `begin` and `end` statements, respectively, in a text are balanced, since this would be, as we shall see in Chapter 8, a variant of the language $\{a^n b^n \mid n \in \mathbb{N}\}$ which is not regular (cf. Theorem 3.2). We therefore continue with a closer look at context-free languages.

6.1 Defining Context-Free Languages

Definition 6.1 *A grammar* $\mathcal{G} = [T, N, \sigma, P]$ *is said to be* context-free (文脈自由文法) *if for all* $\alpha \to \beta$ *in* P *we have* $\alpha \in N$ *and* $\beta \in (N \cup T)^*$.

Definition 6.2 *A language* L *is called* context-free (文脈自由言語) *if there exists a context-free grammar* $\mathcal{G}$ *such that* $L = L(\mathcal{G})$. *By* $\mathcal{CF}$ *we denote the set of all context-free languages.*

Our first theorem shows that the set of all context-free languages is richer than the set of regular languages.

Theorem 6.1 $\mathcal{REG} \subset \mathcal{CF}$.

Proof. Clearly, by definition we have that every regular grammar is also a context-free grammar. Thus, we can conclude $\mathcal{REG} \subseteq \mathcal{CF}$.

For seeing $\mathcal{CF} \setminus \mathcal{REG} \neq \emptyset$ we consider the language $L = \{a^n b^n \mid n \in \mathbb{N}\}$. By Theorem 3.2 we already know $L \notin \mathcal{REG}$. Thus, it suffices to define a context-free grammar $\mathcal{G}$ for L. We set $\mathcal{G} = [\{a, b\}, \{\sigma\}, \sigma, \{\sigma \to a\sigma b, \sigma \to \lambda\}]$. The formal proof of $L = L(\mathcal{G})$ is left as exercise. ∎

Exercise 6.1 *Prove or disprove the language* L *to be context-free, where* $L = \{a^i c^k d^k b^i \mid i, k \in \mathbb{N}^+\}$.

Next, we study closure properties of context-free languages.

6.2 Closure Properties for Context-Free Languages

A first answer is given by showing a theorem analogous to Theorem 2.1.

Theorem 6.2 *The context-free languages are closed under union, product and Kleene closure.*

Proof. Let L_1, L_2 be any context-free languages. Since $L_1, L_2 \in \mathcal{CF}$, there are context-free grammars $\mathcal{G}_1 = [T_1, N_1, \sigma_1, P_1]$ and $\mathcal{G}_2 = [T_2, N_2, \sigma_2, P_2]$ such that $L_i = L(\mathcal{G}_i)$ for $i = 1, 2$. Without loss of generality, we can assume $N_1 \cap N_2 = \emptyset$, for otherwise we rename the nonterminals appropriately.

We start with the union. We have to show that $L = L_1 \cup L_2$ is context-free. This is done in the same way as in the proof of Theorem 2.1.

Next, we deal with the product. We define

$$\mathcal{G}_{prod} = [T_1 \cup T_2, N_1 \cup N_2 \cup \{\sigma\}, \sigma, P_1 \cup P_2 \cup \{\sigma \to \sigma_1 \sigma_2\}].$$

The new production $\sigma \to \sigma_1 \sigma_2$ is context-free. Using *mutatis mutandis* the same ideas as in the proof of Theorem 2.1, $L(\mathcal{G}_{prod}) = L_1 L_2$ follows.

Finally, we deal with Kleene closure. Let $L \in \mathcal{CF}$ and let $\mathcal{G} = [T, N, \sigma, P]$ be a context-free grammar such that $L = L(\mathcal{G})$. We have to show that L^* is context-free. Recall that $L^* = \bigcup_{i \geq 0} L^i$. Since $L^0 = \{\lambda\}$, we have to make sure that λ can be generated. This is obvious if $\lambda \in L$. Otherwise, we simply add the production $\sigma \to \lambda$. Now, it suffices to define

$$\mathcal{G}^* = [T, N \cup \{\sigma^*\}, \sigma^*, P \cup \{\sigma^* \to \sigma\sigma^*, \sigma^* \to \lambda\}] \ .$$

Again, we leave it as an exercise to show that $L(\mathcal{G}^*) = L^*$. ∎

Next, we aim to show $\mathcal{CF}$ to be closed under transposition. For doing it, we introduce the notation $h \stackrel{m}{\Rightarrow} w$ to denote a derivation of length m, i.e., we write $h \stackrel{m}{\Rightarrow} w$ if w can be derived from h within exactly m steps. Also, occasionally we add the subscript $\mathcal{G}$ to $\Rightarrow$, $\stackrel{*}{\Rightarrow}$, or $\stackrel{m}{\Rightarrow}$ to clarify which grammar is used for the generation, e.g., we write $\sigma \stackrel{*}{\underset{\mathcal{G}}{\Longrightarrow}} w$ if w can be generated from σ by using the productions of grammar $\mathcal{G}$.

Additionally, we need the following lemma.

Lemma 6.1 *Let $\mathcal{G} = [T, N, \sigma, P]$ be a context-free grammar and let $\alpha, \beta \in (N \cup T)^*$. If $\alpha \stackrel{m}{\Rightarrow} \beta$ for some $m \geq 0$ and $\alpha = \alpha_1 \cdots \alpha_n$ for some $n \geq 1$, where $\alpha_i \in (N \cup T)^*$ for $i = 1, \ldots, n$, then there exist $t_i \geq 0$, $\beta_i \in (N \cup T)^*$ for $i = 1, \ldots, n$ such that $\beta = \beta_1 \cdots \beta_n$ and $\alpha_i \stackrel{t_i}{\Rightarrow} \beta_i$ and*
$$\sum_{i=1}^{n} t_i = m.$$

Proof. The proof is done by induction on m. For the induction basis we choose $m = 0$ and get $\alpha \stackrel{0}{\Rightarrow} \beta$ implies $\alpha = \beta$. Thus, we can choose $\alpha_i = \beta_i$ as well as $t_i = 0$ for $i = 1, \ldots, n$ and have $\alpha_i \stackrel{0}{\Rightarrow} \beta_i$ for $i = 1, \ldots, n$ and $\sum_{i=1}^{n} t_i = 0$. This proves the induction basis.

Our induction hypothesis is that, if $\alpha = \alpha_1 \cdots \alpha_n \stackrel{m}{\Rightarrow} \beta$, then there exist $t_i \geq 0$, $\beta_i \in (N \cup T)^*$ for $i = 1, \ldots, n$ such that $\beta = \beta_1 \cdots \beta_n$ and $\alpha_i \stackrel{t_i}{\Rightarrow} \beta_i$

and $\sum_{i=1}^{n} t_i = m$.

For the induction step consider $\alpha = \alpha_1 \cdots \alpha_n \overset{m+1}{\Rightarrow} \beta$. Then there exists $\gamma \in (N \cup T)^*$ such that $\alpha = \alpha_1 \cdots \alpha_n \Rightarrow \gamma \overset{m}{\Rightarrow} \beta$.

Since $\mathcal{G}$ is context-free, the production used in $\alpha \Rightarrow \gamma$ must have been of the form $h \to \zeta$, where $h \in N$ and $\zeta \in (N \cup T)^*$. Let α_k, $1 \leq k \leq m$, contain the nonterminal h that was rewritten in using $h \to \zeta$ to obtain $\alpha \Rightarrow \gamma$. Then $\alpha_k = \alpha' h \beta'$ for some $\alpha', \beta' \in (N \cup T)^*$. Furthermore, let

$$\gamma_i = \begin{cases} \alpha_i, & \text{if } i \neq k\,; \\ \alpha' \zeta \beta', & \text{if } i = k\,. \end{cases}$$

Then, for $i \neq k$, $\alpha_i \overset{0}{\Rightarrow} \gamma_i$ and $\alpha_k \Rightarrow \gamma_k$. Thus,

$$\alpha = \alpha_1 \cdots \alpha_n \Rightarrow \gamma_1 \cdots \gamma_n \overset{m}{\Rightarrow} \beta.$$

By the induction hypothesis there exist t_i, β_i such that $\beta = \beta_1 \cdots \beta_n$ and $\gamma_i \overset{t_i}{\Rightarrow} \beta_i$ and $\sum_{i=1}^{n} t_i = m$. Combining the derivations we find

$$\alpha_i \overset{0}{\Rightarrow} \gamma_i \overset{t_i}{\Rightarrow} \beta_i \text{ for } i \neq k \quad \text{and} \quad \alpha_k \Rightarrow \gamma_k \overset{t_k}{\Rightarrow} \beta_k.$$

Furthermore, we set

$$t'_i = \begin{cases} t_i, & \text{if } i \neq k\,; \\ t_k + 1, & \text{if } i = k\,. \end{cases}$$

Thus, we have $t'_i \geq 0$ and $\beta_i \in (N \cup T)^*$ satisfy $\beta = \beta_1 \cdots \beta_n$ and $\alpha_i \overset{t'_i}{\Rightarrow} \beta_i$. Finally, we compute

$$\sum_{i=1}^{n} t'_i = 1 + \sum_{i=1}^{n} t_i = m + 1,$$

and the lemma follows. ∎

6. Context-Free Languages

Now, we are ready to prove $\mathcal{CF}$ to be closed under transposition.

Theorem 6.3 *Let Σ be any alphabet, and let $L \subseteq \Sigma^*$. Then we have: If $L \in \mathcal{CF}$ then $L^T \in \mathcal{CF}$, too.*

Proof. Let L be any context-free language. Hence, there exists a context-free grammar $\mathcal{G} = [T, N, \sigma, P]$ such that $L = L(\mathcal{G})$. Now, we define a grammar $\mathcal{G}^T$ as follows. Let

$$\mathcal{G}^T = [T, N, \sigma, P^T], \text{ where } P^T = \{\alpha^T \to \beta^T \mid (\alpha \to \beta) \in P\}. \quad (6.1)$$

Taking into account that $(\alpha \to \beta) \in P$ implies $\alpha \in N$ and $\beta \in (N \cup T)^*$ (cf. Definition 6.1), and that $\alpha^T = \alpha$ as well as $\beta^T \in (N \cup T)^*$, we can directly conclude that $\mathcal{G}^T$ is context-free. Thus, it remains to show that $L(\mathcal{G}^T) = L^T$. This is done via the following claim.

Claim 1. For each $h \in N$ we have,
$$h \underset{\mathcal{G}}{\overset{m}{\Longrightarrow}} w, \, w \in (N \cup T)^* \text{ if and only if } h \underset{\mathcal{G}^T}{\overset{m}{\Longrightarrow}} w^T.$$

The claim is proved by induction on m. We start with the necessity.

For the induction basis, we choose $m = 0$, and clearly have $h \underset{\mathcal{G}}{\overset{0}{\Longrightarrow}} h$ if and only if $h \underset{\mathcal{G}^T}{\overset{0}{\Longrightarrow}} h$, since $h = h^T$.

Now, we have the following induction hypothesis.

For all $h \in N$, if $h \underset{\mathcal{G}}{\overset{t}{\Longrightarrow}} w$, $w \in (N \cup T)^*$, and $t \le m$, then $h \underset{\mathcal{G}^T}{\overset{t}{\Longrightarrow}} w^T$.

Induction step: We must show that, if $h \underset{\mathcal{G}}{\overset{m+1}{\Longrightarrow}} s$, $s \in (N \cup T)^*$, then $h \underset{\mathcal{G}^T}{\overset{m+1}{\Longrightarrow}} s^T$. Let $h \underset{\mathcal{G}}{\overset{m+1}{\Longrightarrow}} s$, then there is an $\alpha \in (N \cup T)^*$ such that $h \underset{\mathcal{G}}{\Longrightarrow} \alpha \underset{\mathcal{G}}{\overset{m}{\Longrightarrow}} s$. Let

$$\alpha = u_1 h_1 u_2 h_2 \cdots u_n h_n u_{n+1}, \text{ where } u_i \in T^* \text{ and } h_i \in N.$$

Since $h \underset{\mathcal{G}}{\Longrightarrow} \alpha$ implies $(h \to \alpha) \in P$, we get by (6.1) that $(h \to \alpha^T) \in P^T$. Hence, we can conclude $h \underset{\mathcal{G}^T}{\Longrightarrow} \alpha^T$, and obtain

$$h \underset{\mathcal{G}^T}{\Longrightarrow} \alpha^T = u_{n+1}^T h_n u_n^T h_{n-1} \cdots u_2^T h_1 u_1^T.$$

Furthermore, we have

6.2 Closure Properties for Context-Free Languages

$$u_1 h_1 u_2 h_2 \cdots u_n h_n u_{n+1} \underset{\mathcal{G}}{\overset{m}{\Longrightarrow}} s \, ,$$

and therefore, by Lemma 6.1, there exist $\gamma_i \in (N \cup T)^*$ and $t_i \geq 0$ such that $h_i \underset{\mathcal{G}}{\overset{t_i}{\Longrightarrow}} \gamma_i$, as well as

$$s = u_1 \gamma_1 u_2 \gamma_2 \cdots \gamma_n u_{n+1} \, , \quad \text{and} \quad \sum_{i=1}^{n} t_i = m \, .$$

Consequently, $t_i \leq m$ and by the induction hypothesis we obtain that

$$h_i \underset{\mathcal{G}}{\overset{t_i}{\Longrightarrow}} \gamma_i \quad \text{implies} \quad h_i \underset{\mathcal{G}^T}{\overset{t_i}{\Longrightarrow}} \gamma_i^T \, .$$

Therefore,

$$h \underset{\mathcal{G}^T}{\Longrightarrow} u_{n+1}^T h_n u_n^T h_{n-1} \cdots u_2^T h_1 u_1^T \underset{\mathcal{G}^T}{\overset{m}{\Longrightarrow}} u_{n+1}^T \gamma_n^T u_n^T \gamma_{n-1}^T \cdots u_2^T \gamma_1^T u_1^T = s^T \, .$$

This completes the induction step, and thus the necessity is shown.

The sufficiency can be shown analogously by replacing $\mathcal{G}$ by $\mathcal{G}^T$ and $\mathcal{G}^T$ by $\mathcal{G}$. Thus, we obtain $L(\mathcal{G}^T) = L^T$. ∎

Next, we are going to prove that $\mathcal{CF}$ is not closed under intersection.

Theorem 6.4 *There are languages* L_1, $L_2 \in \mathcal{CF}$ *such that* $L_1 \cap L_2 \notin \mathcal{CF}$.

Proof. Let

$L_1 = \{a^n b^n c^m \mid n, m \in \mathbb{N}\}$ and let

$L_2 = \{a^m b^n c^n \mid n, m \in \mathbb{N}\}$.

Then we directly get $L = L_1 \cap L_2 = \{a^n b^n c^n \mid n \in \mathbb{N}\}$, but $L \notin \mathcal{CF}$ as we shall show a bit later, when we have the pumping lemma for context-free languages (cf. Theorem 7.4). ∎

Theorem 6.4 allows for a nice corollary. For stating it, we introduce the following notation. Let $L \subseteq \Sigma^*$ be any language; we use $\overline{L}$ to denote the *complement* (補集合) of L, i.e., $\overline{L} =_{df} \Sigma^* \setminus L$. Now, we are ready for our corollary.

6. Context-Free Languages

Corollary 6.1 $\mathcal{CF}$ *is not closed under complement.*

Proof. The proof is done indirectly. Suppose the converse, i.e., for all context-free languages L_1 and L_2 we also have $\overline{L_1}, \overline{L_2} \in \mathcal{CF}$. By Theorem 6.2 we can conclude $\overline{L_1} \cup \overline{L_2} \in \mathcal{CF}$ (closure under union). But then, by our supposition, we must also have $\overline{\overline{L_1} \cup \overline{L_2}} \in \mathcal{CF}$. Since $\overline{\overline{L_1} \cup \overline{L_2}} = L_1 \cap L_2$ by de Morgan's laws for sets, this would imply the context-free languages to be closed under intersection, a contradiction to Theorem 6.4. ∎

Exercise 6.2 *Prove or disprove: For all* $L \in \mathcal{CF}$ *and* $R \in \mathcal{REG}$ *we have* $L \cap R \in \mathcal{CF}$.

We continue with further properties of context-free languages that are needed later but that are also of independent interest.

First, we start with the following observation. It can sometimes happen that grammars contain nonterminals that cannot generate any terminal string. Let us look at the following example.

Example 6.1 Let $\Sigma = \{a, +, *, (,), -\}$ be the terminal alphabet, and assume the set of productions given is:

$E \to E + E$ $F \to F * (T)$

$E \to E + T$ $F \to a$

$E \to E + F$ $T \to E - T$

$F \to F * E$

where F is the start symbol. Now, we see that the only rule containing T on the left-hand side is $T \to E - T$. Thus it also contains T on the right-hand side. Consequently, from T *no* terminal string can be derived. Such a situation is highly undesirable, since it will only complicate many tasks such as analyzing derivations. Therefore, we introduce the notion of a *reduced grammar* (既約文法).

6.2 Closure Properties for Context-Free Languages

Definition 6.3 *A context-free grammar* $\mathcal{G} = [T, N, \sigma, P]$ *is said to be* reduced (既約) *if*
(1) *for all* $h \in N \setminus \{\sigma\}$ *there are strings* $p, q \in (T \cup N)^*$ *such that* $\sigma \stackrel{*}{\Rightarrow} phq$,
(2) *there is a string* $w \in T^*$ *such that* $h \stackrel{*}{\Rightarrow} w$.

The following theorem shows the usefulness of Definition 6.3.

Theorem 6.5 *For every context-free grammar* $\mathcal{G} = [T, N, \sigma, P]$ *there is a reduced context-free grammar* $\mathcal{G}' = [T', N', \sigma', P']$ *such that* $L(\mathcal{G}) = L(\mathcal{G}')$.

Proof. We execute two steps. First, we collect all $h \in N$ from which a string in T^* can be derived. Let this set be $\tilde{H}$. Second, we collect all those symbols from $\tilde{H}$ that can be reached from the start symbol σ.

For the first step, we set $H_0 = T$ and proceed inductively as follows.

$$H_1 \;\; = H_0 \cup \{h \mid h \in N,\ (h \rightarrow p) \in P \text{ and } p \in H_0^*\},$$
$$H_{i+1} = H_i \cup \{h \mid h \in N,\ (h \rightarrow p) \in P \text{ and } p \in H_i^*\}.$$

By construction we obviously have $H_0 \subseteq H_1 \subseteq H_2 \subseteq \cdots \subseteq T \cup N$.

Since $T \cup N$ is finite, there must be an index i_0 such that $H_{i_0} = H_{i_0+1}$.

Claim 1. If $H_i = H_{i+1}$ then $H_i = H_{i+m}$ for all $m \in \mathbb{N}^+$.

Since $H_{i+1} = H_i$, by construction, the induction basis for $m = 1$ is obvious.

Thus, we have the induction hypothesis $H_i = H_{i+m}$. For the induction step we have to show that $H_i = H_{i+m+1}$.

Let $h \in H_{i+m+1}$, then we either have $h \in H_{i+m}$ (and the induction hypothesis directly applies) or $(h \rightarrow p) \in P$ and $p \in H_{i+m}^*$. By the induction hypothesis we know that $H_i = H_{i+m}$. Hence, the condition $(h \rightarrow p) \in P$ and $p \in H_{i+m}^*$ can be stated as $(h \rightarrow p) \in P$ and $p \in H_i^*$. But the latter condition implies by construction that $h \in H_{i+1}$. By assumption, $H_i = H_{i+1}$, and thus we conclude $h \in H_i$. This proves Claim 1.

Next, we set $\tilde{H} =_{df} H_{i_0} \setminus T$. The following claim holds by construction.

Claim 2. $h \in \tilde{H}$ if and only if there is a $w \in T^*$ such that $h \stackrel{*}{\Rightarrow} w$.

6. Context-Free Languages

Next, we perform the second step. We define inductively

$R_0 = \{\sigma\}$,

$R_1 = R_0 \cup \{h \mid h \in \tilde{H} \wedge \exists p, q[p,q \in (T \cup N)^* \wedge \sigma \overset{1}{\Rightarrow} phq]\}$,

$R_{i+1} = R_i \cup \{h \mid h \in \tilde{H} \wedge \exists r_i \exists p, q[r_i \in R_i, p, q \in (T \cup N)^* \wedge r_i \overset{1}{\Rightarrow} phq]\}$,

for all $i \geq 1$. In the same manner as above one can easily prove that there is an $i_0 \in \mathbb{N}$ such that $R_{i_0} = R_{i_0 + m}$ for all $m \in \mathbb{N}$.

Finally, we define $T' = T$, $N' = R_{i_0} \cap \tilde{H}$, $\sigma' = \sigma$ and

$P' = \emptyset$, if $\sigma \notin N'$;

$P' = P \setminus \{\alpha \to \beta \mid (\alpha \to \beta) \in P,\ \alpha \notin R_{i_0}$ or $\beta \notin (T \cup N')^*\}$, otherwise,

i.e., we get $\mathcal{G}' = [T, N', \sigma, P']$.

By construction $L(\mathcal{G}') = L(\mathcal{G})$. Note that Claim 2 implies Condition (2) of Definition 6.3 and Condition (1) of Definition 6.3 is fulfilled by the definition of the sets R_i. ∎

The proof of the latter theorem directly allows for the following corollary.

Corollary 6.2 *There is an algorithm which, on input any context-free grammar $\mathcal{G}$, decides whether or not $L(\mathcal{G}) = \emptyset$.*

Proof. We use the algorithm from the proof of Theorem 6.5. Let $\mathcal{G}' = [T, N', \sigma, P']$ be the grammar returned. Then we clearly have $L(\mathcal{G}) = \emptyset$ if and only if $\sigma \notin N'$. ∎

Next, we illustrate the construction of a reduced grammar.

Example 6.2 Let $\mathcal{G} = [\{a, b, c, d\}, \{\sigma, \alpha, \beta, \gamma, \delta\}, \sigma, P]$, where P is defined as follows.

$\sigma \to \alpha\beta$	$\beta \to \gamma\beta$	$\gamma \to b$
$\sigma \to \gamma\alpha$	$\beta \to \alpha\beta$	$\delta \to a\delta$
$\alpha \to a$	$\gamma \to c\beta$	$\delta \to d$

6.2 Closure Properties for Context-Free Languages

So, we obtain:

$H_0 = \{a, b, c, d\}$,

$H_1 = \{a, b, c, d, \alpha, \gamma, \delta\}$,

$H_2 = \{a, b, c, d, \alpha, \gamma, \delta, \sigma\}$,

$H_3 = \{a, b, c, d, \alpha, \gamma, \delta, \sigma\} = H_2$,

and thus we terminate. Consequently, $\tilde{H} = \{\alpha, \gamma, \delta, \sigma\}$.

Next, we compute the sets R_i, i.e., $R_0 = \{\sigma\}$, $R_1 = \{\sigma, \alpha, \gamma\} = R_2$, and we terminate again. Thus, $N' = \{\sigma, \alpha, \gamma\}$.

Finally, we delete all productions having in their left hand side a symbol not in $\{\sigma, \alpha, \gamma\}$, i.e., $\beta \to \gamma\beta$, $\beta \to \alpha\beta$, $\delta \to a\delta$, and $\delta \to d$. Then we delete from the remaining productions those containing a symbol in their right hand side not in N', i.e., $\sigma \to \alpha\beta$ and $\gamma \to c\beta$. Thus, we obtain the grammar $\mathcal{G}' = [\{a, b, c, d\}, \{\sigma, \alpha, \gamma\}, \sigma, \{\sigma \to \gamma\alpha, \alpha \to a, \gamma \to b\}]$.

So, we also see that $\sigma \in N'$. Therefore, by Corollary 6.2, we conclude that $L(\mathcal{G}) \neq \emptyset$.

Example 6.3 Let $\mathcal{G} = [\{a\}, \{\sigma, \beta, \gamma\}, \sigma, P]$, where

$$P = \{\sigma \to \beta\gamma, \beta \to a, \gamma \to \gamma\}.$$

Then we obtain, $H_0 = \{a\}$, $H_1 = \{a, \beta\} = H_2$, and we terminate. Thus $\tilde{H} = \{\beta\}$. Next, we get $R_0 = \{\sigma\}$ and $R_1 = \{\sigma, \beta\} = R_2$. Thus, we terminate again. This is the point where we need the intersection in the definition of N', since now $N' = \{\beta\}$. Consequently, in the definition of P' the first case applies and $P' = \emptyset$. Also, $\sigma \notin N'$ and thus $L(\mathcal{G}) = \emptyset$.

We continue with the following definition.

Definition 6.4 *A grammar* $\mathcal{G} = [T, N, \sigma, P]$ *is said to be* λ-free (λ-自由) *if P does not contain any production of the form* $h \to \lambda$.

Theorem 6.6 *For every context-free grammar* $\mathcal{G} = [T, N, \sigma, P]$ *there is a context-free grammar* $\mathcal{G}'$ *such that* $L(\mathcal{G}') = L(\mathcal{G}) \setminus \{\lambda\}$ *and* $\mathcal{G}'$ *is* λ-*free.*

Furthermore, if $\lambda \in L(\mathcal{G})$ *then there exists an equivalent context-free grammar* $\mathcal{G}''$ *such that* $\sigma'' \to \lambda$ *is the only production having* λ *on its right-hand side and* σ'' *does not occur at any right-hand side.*

We leave it as an exercise to prove this theorem.

6.3 Problem Set 6

Problem 6.1 Let $\mathcal{G} = [\{a, +, *, (,), -\}, \{F, E, T\}, F, P]$, where P is

$E \to E + E$ $\qquad$ $F \to F * (T)$

$E \to E + T$ $\qquad$ $F \to a$

$E \to E + F$ $\qquad$ $T \to E - T$

$F \to F * E$

Construct a reduced grammar $\mathcal{G}'$ such that $L(\mathcal{G}) = L(\mathcal{G}')$.

Problem 6.2 Construct context-free grammars for the following languages:

(1) $L = \{a^{3i}b^i \mid i \in \mathbb{N}^+\}$,

(2) $L = \{a^i b^j \mid i, j \in \mathbb{N}^+ \text{ and } i \geq j\}$,

(3) $L = \{s \mid s \in \{a, b\}^* \text{ and the number of } a's \text{ in } s$ equals the number of $b's$ in $s\}$.

Problem 6.3 Prove Theorem 6.6.

7 More About Context-Free Languages

We start this chapter by defining the so-called BNF. Then we take a look at parse trees. Furthermore, we shall define the first normal form for context-free grammars, i.e., the Chomsky normal form. Then we state the pumping lemma for context-free languages, and discuss some consequences.

7.1 Backus-Naur Form

Context-free grammars are of fundamental importance for programming languages. In the specification of programming languages usually a form different to the one provided in Definition 6.1 is used. This form is the so-called *Backus normal form* (バッカス標準形) or *Backus-Naur form* (バッカス-ナウア形). It was created by John Backus to specify the grammar of ALGOL. Later it was simplified by Peter Naur to reduce the character set used and Donald E. Knuth proposed to call the new form Backus-Naur form. Fortunately, whether or not one follows Knuth's suggestion, the form is commonly abbreviated by BNF.

The form uses four meta characters that are not allowed to appear in the working vocabulary, i.e., in $T \cup N$. These meta characters are < > ::= | and they are used as follows. Strings (*not* containing the meta characters) are enclosed by < and > denote nonterminals. The symbol ::= serves as replacement operator (in the same way as →) and | is read as "or."

The example below is from Harrison [12]. Consider an ordinary context-free grammar for unsigned integers in a programming language. Here, D stands for the class of digits and U for the class of unsigned integers.

Example 7.1

D → 0	D → 4	D → 8
D → 1	D → 5	D → 9
D → 2	D → 6	U → D
D → 3	D → 7	U → UD

Rewriting this example in BNF yields:

<digit>   ::= 0|1|2|3|4|5|6|7|8|9

<unsigned integer> ::= <digit>|<unsigned integer><digit> .

This example shows that the BNF allows for a very compact representation of the grammar. This is important when defining the syntax of a programming language, where the set of productions is usually large.

Whenever appropriate, we shall adopt a blend of the notation used in BNFs, i.e., occasionally we shall use | as well as < and > but not ::= .

Context-free languages play an important role in many applications. As far as regular languages are concerned, we have seen that finite automata are very efficient recognizers. So, what about context-free languages? Again, for every context-free language a recognizer can be algorithmically constructed. Formally, these recognizers are *pushdown automata* (プッシュダウンオートマトン). There are many software systems around that perform the construction of the relevant pushdown automaton for a given language. These systems are important in that they allow the quick construction of the syntax analysis part of a compiler for a new language and are therefore highly valued. We shall come back to this topic later (see Chapters 9 and 10), since it is better to treat another important tool for syntax analysis first, i.e., *parsers* (構文解析器, パーサ).

One of the most widely used of these syntax analyzer generators is called yacc (yet another compiler-compiler). The generation of a parser, i.e., a function that creates parse trees from source programs has been institutionalized in the yacc command that appears in all UNIX systems.

The input to yacc is a context-free grammar, in a notation that differs only in details from the BNF. Associated with each production is an *action* (アクション), which is a fragment of C code that is performed whenever a node of the parse tree that (combined with its children) corresponds to this production is created.

7.2 Parse Trees, Ambiguity

A nice feature of grammars is that they describe the hierarchical syntactic structure of the sentences of languages they define. These hierarchical structures are described by *parse trees* (構文木).

Parse trees are a representation for derivations. When used in a compiler, it is the data structure of choice to represent the source program. In a compiler, the tree structure of the source program facilitates the translation of the source program into executable code by allowing natural, recursive functions to perform this translation process.

Definition 7.1 (Parse Tree) *Let* $\mathcal{G} = [T, N, \sigma, P]$ *be a context-free grammar. A* parse tree *for* $\mathcal{G}$ *is a tree satisfying the following conditions.*

(1) *Each interior node and the root are labeled by a variable from* N.

(2) *Each leaf is either labeled by a non-terminal, a terminal or the empty*

string. If the leaf is labeled by the empty string, then it must be the only child of its parent.

(3) If an interior node is labeled by h and its children are labeled by $x_1, \ldots, x_k$, respectively, from left to right, then h $\to$ $x_1 \cdots x_k$ is a production from P.

Thus, every subtree of a parse tree describes one instance of an abstraction in the statement. Next, we define the yield of a parse tree. If we look at the leaves of any parse tree and concatenate them from left to right, we get a string. This string is called the *yield* (成果) of the parse tree.

Exercise 7.1 *Prove that the yield is always a string that is derivable from the start symbol provided the root is labeled by* σ.

Of special importance is the case that the root is labeled by the start symbol and that the yield is a *terminal string* (終端文字列), i.e., all leaves are labeled with a symbol from T or the empty string. Thus, the language of a grammar can also be expressed as the set of yields of those parse trees having the start symbol at the root and a terminal string as yield.

We continue with an example. Assume that we have the following part of a grammar on hand describing how assignment statements are generated.

Example 7.2

```
<assign>  → <id> := <expr>
<id>      → A | B | C
<expr>    → <id> + <expr>
          | <id> * <expr>
          | (<expr>)
          | <id>
```

Now, let us look at the assignment statement A := B * (A + C) which can be generated by the following derivation.

$$\begin{aligned}
\text{<assign>} &\Rightarrow \text{<id>} := \text{<expr>} \\
&\Rightarrow \text{A} := \text{<expr>} \\
&\Rightarrow \text{A} := \text{<id>} * \text{<expr>} \\
&\Rightarrow \text{A} := \text{B} * \text{<expr>} \\
&\Rightarrow \text{A} := \text{B} * (\text{<expr>}) \\
&\Rightarrow \text{A} := \text{B} * (\text{<id>} + \text{<expr>}) \\
&\Rightarrow \text{A} := \text{B} * (\text{A} + \text{<expr>}) \\
&\Rightarrow \text{A} := \text{B} * (\text{A} + \text{<id>}) \\
&\Rightarrow \text{A} := \text{B} * (\text{A} + \text{C})
\end{aligned}$$

The structure of the assignment statement that we have just derived is shown in the parse tree displayed in Figure 7.1.

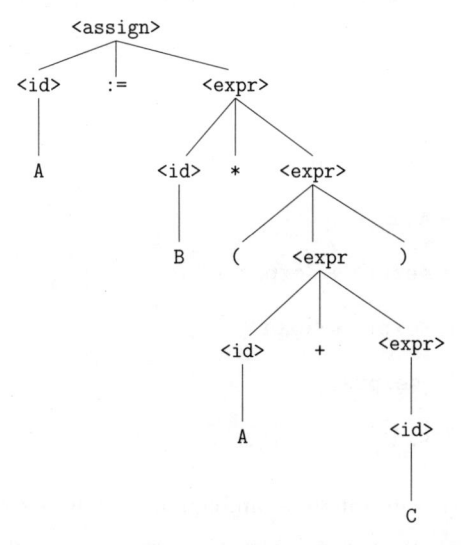

Figure 7.1 Parse tree for $\text{A} := \text{B} * (\text{A} + \text{C})$.

Note that syntax analyzers for programming languages, which are often called *parsers*, construct parse trees for given programs. Some systems construct parse trees only implicitly, but they also use the whole information provided by the parse tree during the parse. There are two major approaches to build these parse trees. One is top-down and the other one is bottom-up. In the top-down approach the parse tree is built from the root to the leaves while in the bottom-up approach the parse tree is built from the leaves upward to the root. A major problem one has to handle when constructing such parsers is *ambiguity* (曖昧さ).

A grammar that generates a sentence for which there are two or more distinct parse trees is said to be *ambiguous* (曖昧な). For having an example, let us look at the following part of a grammar given in Example 7.3. At first glance this grammar looks quite similar to the one considered above. The only difference is that the production for expressions has been altered by replacing <id> by <expr>. However, this small modification leads to serious problems, because now the grammar provides slightly less syntactic structure than the grammar considered in Example 7.2 does.

Example 7.3

<assign> → <id> := <expr>

<id>   → A | B | C

<expr> → <expr> + <expr>

      | <expr> ∗ <expr>

      | (<expr>)

      | <id>

For seeing that this grammar is ambiguous, let us look at the following assignment statement: A := B + C ∗ A .

7.2 Parse Trees, Ambiguity   65

We skip the two formal derivations possible for this assignment and look directly at the two parse trees. These two distinct parse trees cause problems because compilers base the semantics of the sentences on their syntactic structure. In particular, compilers decide what code to generate by examining the parse tree. So, in our example the *semantics* is not clear. Let us examine this problem in some more detail.

In the first parse tree (the left one) of Figure 7.2 the multiplication operator is generated lower in the tree which would indicate that it has precedence over the addition operator in the expression. The second parse tree in Figure 7.2, however, is just indicating the opposite. Clearly, in dependence on what decision the compiler makes, the result of an actual evaluation of the assignment given will be either the expected one (that is multiplication has precedence over addition) or an erroneous one.

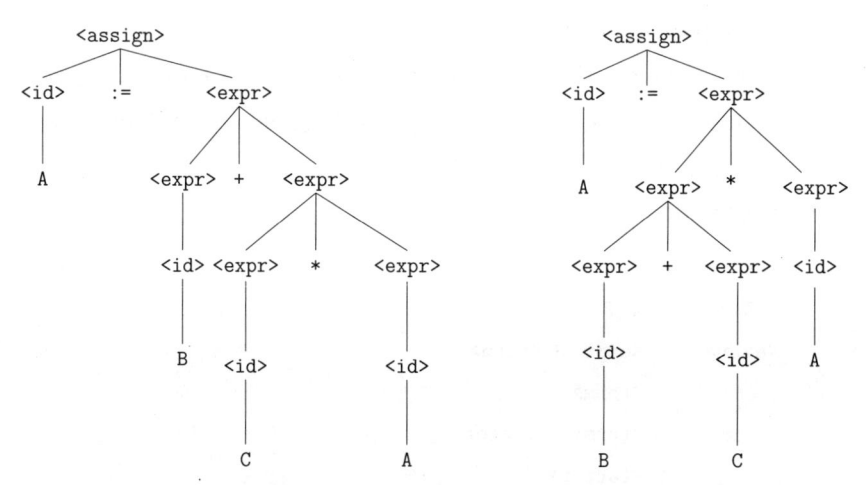

Figure 7.2 Two parse trees for the same sentence A := B+C*A.

Although the grammar in Example 7.2 is not ambiguous, the precedence order of its operators is not the usual one. Rather, in this grammar, a parse tree of a sentence with multiple operators has the rightmost operator at the lowest point, with the other operators in the tree moving progressively higher as one moves to the left in the expression.

We aim to resolve this problem and to clearly define the usual operator precedence between multiplication and addition, or more generally with any desired operator precedence. We achieve this goal for our example by using separate nonterminals for the operands of the operators that have different precedence. This requires additional nonterminals and productions.

However, in general, the situation is much more subtle. First, there is *no* algorithm deciding whether or not any context-free grammar is ambiguous. Second, there are context-free languages that have nothing but ambiguous grammars. See Section 5.4.4 in [15] for a detailed treatment of this issue. But in practice the situation is not as grim as it may seem. Many techniques have been proposed to eliminate ambiguity in the sorts of constructs that typically appear in programming languages.

Next, we eliminate the ambiguity detected in our example. The grammar below generates the same language as the grammars of Examples 7.2 and 7.3 and gives the usual precedence order of multiplication and addition.

Example 7.4

```
<assign>  →  <id> := <expr>
<id>      →  A | B | C
<expr>    →  <expr> + <term>
          |  <term>
<term>    →  <term> * <factor>
          |  <factor>
<factor>  →  ( <expr> )
          |  <id>
```

Let us derive the same statement as above, i.e., $A := B + C * A$. The derivation is unambiguously obtained as follows.

7.2 Parse Trees, Ambiguity

$$\begin{aligned}
\texttt{<assign>} &\Rightarrow \texttt{<id>} := \texttt{<expr>} \\
&\Rightarrow \texttt{A} := \texttt{<expr>} \\
&\Rightarrow \texttt{A} := \texttt{<expr>} + \texttt{<term>} \\
&\Rightarrow \texttt{A} := \texttt{<term>} + \texttt{<term>} \\
&\Rightarrow \texttt{A} := \texttt{<factor>} + \texttt{<term>} \\
&\Rightarrow \texttt{A} := \texttt{<id>} + \texttt{<term>} \\
&\Rightarrow \texttt{A} := \texttt{B} + \texttt{<term>} \\
&\Rightarrow \texttt{A} := \texttt{B} + \texttt{<term>} * \texttt{<factor>} \\
&\Rightarrow \texttt{A} := \texttt{B} + \texttt{<factor>} * \texttt{<factor>} \\
&\Rightarrow \texttt{A} := \texttt{B} + \texttt{<id>} * \texttt{<factor>} \\
&\Rightarrow \texttt{A} := \texttt{B} + \texttt{C} * \texttt{<factor>} \\
&\Rightarrow \texttt{A} := \texttt{B} + \texttt{C} * \texttt{<id>} \\
&\Rightarrow \texttt{A} := \texttt{B} + \texttt{C} * \texttt{A}
\end{aligned}$$

Exercise 7.2 *Construct the parse tree for this sentence in accordance with the derivation given above.*

We should note that we have presented a *leftmost derivation* (最左導出) above, i.e., the leftmost nonterminal has always been handled first until it was replaced by a terminal. Thus, the sentence given above has more than one derivation. If we had always handled the rightmost nonterminal first, then we would have been arrived at a *rightmost derivation* (最右導出).

Exercise 7.3 *Give a rightmost derivation of the sentence* $\texttt{A} := \texttt{B}+\texttt{C}*\texttt{A}$.

Exercise 7.4 *Construct the parse tree corresponding to the rightmost derivation and compare it to the parse tree obtained in Exercise 7.2.*

Clearly, one could also choose arbitrarily the nonterminal which allows the application of a production. Try it out, and construct the resulting parse trees. Usually, one is implicitly assuming that derivations are leftmost. Thus, we can more precisely say that a context-free grammar is ambiguous if there is a sentence in the language it generates that possesses at least two different leftmost derivations. Otherwise it is called *unambiguous* (無曖昧). A language is said to be unambiguous if there is an unambiguous grammar for it.

Exercise 7.5 *Prove or disprove that every* $L \in \mathcal{REG}$ *is unambiguous.*

Another important problem in describing programming languages is to express that operators are *associative*. As we have learned in mathematics, addition and multiplication are associative. Is this also true for computer arithmetic? As far as integer addition and multiplication are concerned, they are associative. But floating point computer arithmetic is not always associative. So, in general correct associativity is essential.

For example, look at the expression $A := B + C + A$.
Then it should not matter whether or not the expression is evaluated in left order (that is $(B + C) + A$) or in right order (i.e., $B + (C + A)$).

Some programming languages define in what order expressions are evaluated. In particular, in most programming languages that provide it, the exponentiation operator is right associative. The formal tool to describe right (or left) associativity is right (or left) recursion in the corresponding productions. If a BNF production has its left hand side also appearing at the beginning of its right hand side, then the rule is said to be *left recursive* (左再帰). Analogously, if a BNF rule has its left hand side also appearing at the right end of its right hand side, then it is *right recursive* (右再帰).

The following grammar exemplifies of how to use right recursiveness to express right associativity of the exponentiation operator.

```
<factor> → <exp> * *<factor>
         | <exp>
<exp>    → <expr>
         | <id>
```

Finally, we should have a look at the **if-then-else** statement that is present in many programming languages. Is there an unambiguous grammar for it? This is indeed the case, but we leave it as an exercise.

7.3 Chomsky Normal Form

In a context-free grammar, there is no *a priori* bound on the size of a right-hand side of a production. This may complicate many proofs. Fortunately, there is a normal form for context-free grammars bounding the right-hand side to be of length at most 2. Knowing and applying this considerably simplifies many proofs.

First, we define the notion of a *separated grammar* （分離的文法）.

Definition 7.2 *A grammar* $\mathcal{G} = [T, N, \sigma, P]$ *is called* separated （分離的文法） *if either* $\alpha, \beta \in N^*$ *or* $\alpha \in N$ *and* $\beta \in T$ *for all* $(\alpha \to \beta) \in P$.

Theorem 7.1 *For every context-free grammar* $\mathcal{G} = [T, N, \sigma, P]$ *there exists an equivalent separated context-free grammar* $\mathcal{G}' = [T, N', \sigma, P']$.

Proof. First, we introduce for every $t \in T$ a new nonterminal symbol h_t, where new means that $h_t \notin N$ for all $t \in T$. We set $N' = \{h_t \mid t \in T\} \cup N$.

Next, for $(\alpha \to \beta) \in P$, we denote the production obtained by replacing every terminal symbol t in β by h_t by $(\alpha \to \beta)[t/\!/h_t]$. The production set P' is then defined as follows:

$$P' = \{(\alpha \to \beta)[t/\!/h_t] \mid (\alpha \to \beta) \in P\} \cup \{h_t \to t \mid t \in T\}.$$

By construction, we directly see that $\mathcal{G}'$ is separated. Also, the construction ensures that $\mathcal{G}'$ is context-free. It remains to show that $L(\mathcal{G}) = L(\mathcal{G}')$.

Claim 1. $L(\mathcal{G}) \subseteq L(\mathcal{G}')$.

Let $s \in L(\mathcal{G})$. Then there exists a derivation

$$\sigma \stackrel{1}{\Rightarrow} w_1 \stackrel{1}{\Rightarrow} w_2 \stackrel{1}{\Rightarrow} \cdots \stackrel{1}{\Rightarrow} w_n \stackrel{1}{\Rightarrow} s \;,$$

where $w_1, \ldots, w_n \in (N \cup T)^+ \setminus T^*$, and $s \in T^*$. Let P_i be the production used to generate w_i, $i = 1, \ldots, n$.

Then we can generate s by using productions from P' as follows. Let $s = s_1 \cdots s_m$, where $s_j \in T$ for all $j = 1, \ldots, m$. Instead of applying P_i we use $P_i[t /\!\!/ h_t]$ from P' and obtain

$$\sigma \stackrel{1}{\Rightarrow} w'_1 \stackrel{1}{\Rightarrow} w'_2 \stackrel{1}{\Rightarrow} \cdots \stackrel{1}{\Rightarrow} w'_n \stackrel{1}{\Rightarrow} h_{s_1} \cdots h_{s_m} \;,$$

where now we have $w'_i \in (N')^+$ for all $i = 1, \ldots, n$.

Thus, for obtaining s, now it suffices to apply the productions $h_{s_j} \to s_j$ for $j = 1, \ldots m$. This proves Claim 1.

Claim 2. $L(\mathcal{G}') \subseteq L(\mathcal{G})$.

This claim can be proved analogously by inverting the construction used in showing Claim 1. ∎

Now, we are ready to define the Chomsky normal form （チョムスキー標準形）, abbr. CNF, for context-free grammars announced above.

Definition 7.3 (Chomsky Normal Form) *A grammar* $\mathcal{G} = [T, N, \sigma, P]$ *is said to be in* Chomsky normal form *if all productions of* P *have the form* $h \to h_1 h_2$, *where* $h, h_1, h_2 \in N$, *or* $h \to x$, *where* $h \in N$ *and* $x \in T$.

The latter definition directly allows for the following corollary.

Corollary 7.1 *Let* $\mathcal{G} = [T, N, \sigma, P]$ *be a grammar in CNF. Then we have*

(1) $\mathcal{G}$ *is context-free,*

(2) $\mathcal{G}$ *is* λ-*free,*

(3) $\mathcal{G}$ *is separated.*

7.3 Chomsky Normal Form

The following theorem shows the usefulness of Definition 7.3.

Theorem 7.2 *For every context-free grammar* $\mathcal{G} = [T, N, \sigma, P]$ *such that* $\lambda \notin L(\mathcal{G})$ *there exists an equivalent grammar* $\mathcal{G}'$ *that is in CNF.*

Proof. Let $\mathcal{G} = [T, N, \sigma, P]$ be given. Without loss of generality, we may assume that $\mathcal{G}$ is reduced (cf. Theorem 6.5). First, we eliminate all productions of the form $h \to h'$. This is done as follows. We set

$$W_0(h) = \{h\} \quad \text{for every } h \in N, \text{ and for each } i \geq 0 \text{ we define}$$
$$W_{i+1}(h) = W_i \cup \{\tilde{h} \mid \tilde{h} \in N \text{ and } (\hat{h} \to \tilde{h}) \in P \text{ for some } \hat{h} \in W_i(h)\}.$$

Then, using the techniques developed so far, the following facts are obvious:

(1) $W_i(h) \subseteq W_{i+1}(h)$ for all $i \geq 0$,
(2) If $W_i(h) = W_{i+1}(h)$ then $W_i(h) = W_{i+m}(h)$ for all $m \in \mathbb{N}$,
(3) $W_n(h) = W_{n+1}(h)$ for $n = \text{card}(N)$,
(4) $W_n(h) = \{B \mid B \in N \text{ and } h \stackrel{*}{\Rightarrow} B\}$.

Now, we define

$$P_1 = \{h \to \gamma \mid h \in N \land \gamma \notin N \land (B \to \gamma) \in P \text{ for some } B \in W_n(h)\}.$$

Let $\mathcal{G}_1 = [T, N, \sigma, P_1]$, then by construction, P_1 does not contain any production of the form $h \to h'$. These productions have been replaced by $h \to \gamma$. That is, if we had $h \stackrel{*}{\Rightarrow} B$ by using the productions from P and $B \to \gamma$, then we now have the production $h \to \gamma$ in P_1. Also note that P_1 contains all original productions $(h \to \gamma) \in P$, where $\gamma \notin N$ by the definition of W_0. The formal verification of $L(\mathcal{G}_1) = L(\mathcal{G})$ is left as exercise.

Next, from $\mathcal{G}_1$ we construct an equivalent separated grammar $\mathcal{G}_2$ by using the algorithm given in the proof of Theorem 7.1. Now, the only productions in P_2 that still need modification are of the form

$$h \to h_1 h_2 \cdots h_n, \quad \text{where } n \geq 3.$$

7. More About Context-Free Languages

We replace any such production by the following productions

$$h \to h_1 h_{h_2 \cdots h_n}$$
$$h_{h_2 \cdots h_n} \to h_2 h_{h_3 \cdots h_n}$$
.
.
.
$$h_{h_{n-1} h_n} \to h_{n-1} h_n \ .$$

Hence, the resulting grammar $\mathcal{G}'$ is in CNF and by construction equivalent to $\mathcal{G}$. We omit the details. ∎

Exercise 7.6 *Let* $\mathcal{G} = [\{a, b, c\}, \{\sigma, h\}, \sigma, P]$, *where the set* P *of productions is* $P = \{\sigma \to hbh,\ h \to hah,\ h \to ca\}$. *Construct a grammar* $\mathcal{G}'$ *which is in CNF and which is equivalent to grammar* $\mathcal{G}$.

Exercise 7.7 *Extend the notion of CNF to context-free languages containing* λ.

We finish this chapter by pointing to an important result which can be proved by using the CNF. This result is usually referred to as Pumping Lemma for context-free languages or Lemma of Bar-Hillel (バー・ヒレル の補題) or qrsuv-Theorem. The reader should compare this theorem to the Pumping Lemma for regular languages (see Lemma 4.2). Note that Theorem 7.3 provides a necessary condition for a language to be context-free. It is not sufficient. So its main importance lies in the fact that one can often use it to show that a language is *not* context-free.

Theorem 7.3 *For every language* $L \in \mathcal{CF}$ *there is a number* k *such that for every* $w \in L$ *with* $|w| \geq k$ *there are strings* q, r, s, u, v *such that*

(1) $w = qrsuv$,

(2) $|rsu| \leq k$,

(3) $ru \neq \lambda$, *and*

(4) $qr^i su^i v \in L$ *for all* $i \in \mathbb{N}$.

7.3 Chomsky Normal Form

Proof. Let $\mathcal{G} = [T, N, \sigma, P]$ be any context-free grammar in CNF for L. Then all parse trees are binary trees except the last derivation step. Let $n = \text{card}(N)$ and $k = 2^n$ and consider the parse tree for any string $w \in L(\mathcal{G})$ with $|w| \geq k$. So, any of the parse trees for w must have depth at least n. Hence, there must exist a path from the root σ to a leaf having length at least n. Including σ there are $n + 1$ nonterminals on this path. Hence, some nonterminal h has to appear at least twice (cf. Figure 7.3, Part (a)).

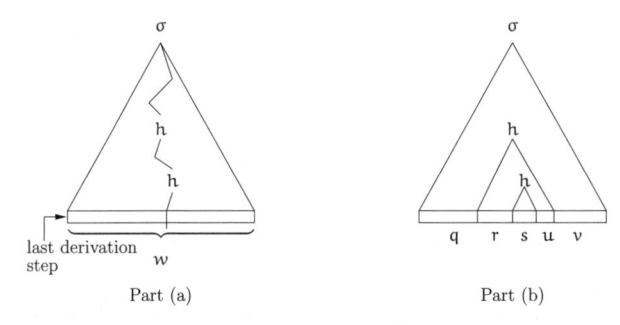

Figure 7.3 Parse trees for illustrating the proof of the qrsuv-Theorem.

Now, starting from the leaves and going up, we fix a path containing two times h and we fix the first two occurrences of h. In this way, we guarantee that the higher located occurrence of h is at most n steps above the leaves.

Next, we look at the substrings that are generated from these two occurrences of h. This gives us the substrings qrsuv (cf. Figure 7.3, Part (b)).

Because $\mathcal{G}$ is in CNF, at the higher occurrence of h a production of the form $h \to h_1 h_2$ must have been applied. Consequently, $r \neq \lambda$ or $u \neq \lambda$, and Assertion (3) is shown.

Assertion (2) follows from the fact that the higher located occurrence of h is at most n steps above the leaves. Thus, the string rsu derived from this h can have length at most $2^n = k$.

It remains to show Assertion (4). This is done by modifying the parse tree. First, we may remove the subtree rooted at the higher located occurrence of h and replace it by the subtree rooted at the lower located occurrence of h (cf. Figure 7.4, Part (a)). In this way, we get a generation of qsv, i.e., $qr^0su^0v \in L(\mathcal{G})$.

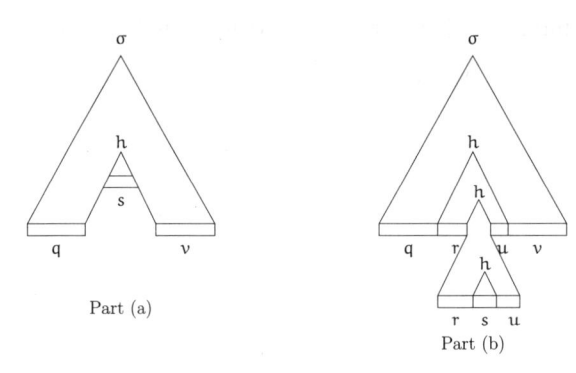

Figure 7.4 Parse trees for illustrating Assertion (4) of the $qrsuv$-Theorem.

Second, we remove the subtree rooted at the lower located occurrence of h and replace it by the subtree rooted at the higher located occurrence of h (cf. Figure 7.4, Part (b)). In this way, we get a derivation of qr^2su^2v and thus, $qr^2su^2v \in L(\mathcal{G})$. Iterating this idea shows $qr^isu^iv \in L(\mathcal{G})$ for every $i \in \mathbb{N}^+$. ∎

For having an application of Theorem 7.3, we show the following Theorem, thus completing the proof of Theorem 6.4.

Theorem 7.4 $L = \{a^nb^nc^n \mid n \in \mathbb{N}\} \notin \mathcal{CF}$.

Proof. Suppose the converse. Then, by Theorem 7.3 there is a k such that for all $w \in L$ with $|w| \geq k$ we have the following.

There are substrings q, r, s, u, v with $|ru| > 0$ and for all $w = qrsuv \in L$ we must have $qr^isu^iv \in L$, too, for all $i \in \mathbb{N}$. Now, consider $w = a^kb^kc^k$. Consequently, $|w| = 3k > k$. By Property (2) of Theorem 7.3 we have

$|rsu| \leq k$, and thus ru cannot contain a's, b's *and* c's. Since $|ru| > 0$ and by Property (4) (applied for $i = 0$), we must have $qr^0su^0v = qsv \in L$. But, as discussed above, $qsv \neq a^\ell b^\ell c^\ell$ for every $\ell \in \mathbb{N}$, a contradiction. ∎

We recommend to solve the following exercises.

Exercise 7.8 *Consider* $L = \{a^p \mid p \text{ is prime}\}$. *Prove or disprove* $L \in \mathcal{CF}$.

Exercise 7.9 *Consider* $L = \{a^{2^n} \mid n \in \mathbb{N}\}$. *Prove or disprove* $L \in \mathcal{CF}$.

Exercise 7.10 *Let* $L_{pal2} = \{ww^T \mid w \in \{0,1\}^*\}$ *be the language of all palindromes of even length over* $\{0, 1\}$. *Show* L_{pal2} *to be context-free and provide a grammar* $\mathcal{G}$ *in CNF such that* $L(\mathcal{G}) = L_{pal2}$.

Exercise 7.11 *Let* Σ *be any alphabet such that* $\text{card}(\Sigma) = 1$. *Prove that*

$$\mathcal{CS} \cap \wp(\Sigma^*) = \mathcal{REG} \cap \wp(\Sigma^*),$$

i.e., every context-free language $L \subseteq \Sigma^*$ *is regular.*

Finally, we provide the problem set for this chapter.

7.4 Problem Set 7

Problem 7.1 Prove or disprove the following theorem.

Theorem 7.5 *There is an algorithm which, on input any context-free grammar* $\mathcal{G}$, *decides whether or not* $\lambda \in L(\mathcal{G})$.

Problem 7.2 We already know that $\mathcal{REG}$ is closed under intersection. In this problem we turn our attention to the algorithmic aspect.

(1) Let $\mathcal{G}$ and $\tilde{\mathcal{G}}$ be any two regular grammars. Construct a regular grammar $\mathcal{G}_\cap$ such that $L(\mathcal{G}_\cap) = L(\mathcal{G}) \cap L(\tilde{\mathcal{G}})$.

(2) Let $\mathcal{A}$ and $\tilde{\mathcal{A}}$ be any two DFAs. Construct a DFA $\mathcal{A}_\cap$ such that $L(\mathcal{A}_\cap) = L(\mathcal{A}) \cap L(\tilde{\mathcal{A}})$.

Problem 7.3 Consider $L = \{(ab)^{3^n} \mid n \in \mathbb{N}\}$. Prove or disprove L to be context-free.

8 | CF and Homomorphisms

In this chapter we study further useful properties and characterizations of context-free languages. First, we look at substitutions.

8.1 Substitutions and Homomorphisms

Since we aim to prove theorems, some formal definitions and notations are needed. Recall that we write $\wp(X)$ to denote the power set of X.

Definition 8.1 *Let Σ and Δ be any two finite alphabets. A mapping $\tau: \Sigma \longrightarrow \wp(\Delta^*)$ is said to be a* substitution (代入). *We extend τ to be a mapping $\tau: \Sigma^* \longrightarrow \wp(\Delta^*)$ (i.e., to strings) by defining*

(1) $\tau(\lambda) = \lambda$,

(2) $\tau(wx) = \tau(w)\tau(x)$ *for all $w \in \Sigma^*$ and $x \in \Sigma$.*

We generalize τ to languages $L \subseteq \Sigma^*$ by setting $\tau(L) = \bigcup_{w \in L} \tau(w)$.

So, a substitution maps every symbol of Σ to a language over Δ. The language a symbol is mapped to can be finite or infinite.

Example 8.1 Let $\Sigma = \{0, 1\}$ and let $\Delta = \{a, b\}$. Then, the mapping τ defined by $\tau(\lambda) = \lambda$, $\tau(0) = \{a\}$ and $\tau(1) = \{b\}^*$ is a substitution.

Let us calculate $\tau(010)$. By definition,

$$\tau(010) = \tau(01)\tau(0) = \tau(0)\tau(1)\tau(0) = \{a\}\{b\}^*\{a\} = \underline{a}\underline{\langle b \rangle}\underline{a} ,$$

where the latter equality is by the definition of regular expressions.

8.1 Substitutions and Homomorphisms

Next, we want to define what is meant by closure of a language family $\mathcal{L}$ under substitutions. Here special care is necessary. At first glance, we may be tempted to require that for every substitution τ the condition $\tau(L) \in \mathcal{L}$ has to be satisfied. But this is a *too strong demand*. Consider $\Sigma = \{0, 1\}$, $\Delta = \{a, b\}$ and $\mathcal{L} = \mathcal{REG}$. Furthermore, suppose that $\tau(0) = L$, where L is any language $L \in \mathcal{CF} \setminus \mathcal{REG}$ over Δ. Then we obviously have $\tau(\{0\}) = L$, too. Consequently, $\tau(\{0\}) \notin \mathcal{REG}$. On the other hand, $\{0\} \in \mathcal{REG}$, and thus we would conclude that $\mathcal{REG}$ is not closed under substitution. A similar argument would prove that $\mathcal{CF}$ is not closed under substitution.

The point to be made here is that we have to restrict the set of allowed substitutions to those ones that map the elements of Σ to languages belonging to $\mathcal{L}$. Therefore, we arrive at the following definition.

Definition 8.2 *Let Σ be any alphabet, and let $\mathcal{L}$ be any language family over Σ. We say that $\mathcal{L}$ is* closed under substitutions *if for every substitution* $\tau \colon \Sigma \longrightarrow \mathcal{L}$ *and every $L \in \mathcal{L}$ we have $\tau(L) \in \mathcal{L}$.*

A very interesting special case of a substitution is the homomorphism.

Definition 8.3 *Let Σ and Δ be any two finite alphabets. A mapping $\varphi \colon \Sigma^* \longrightarrow \Delta^*$ is said to be a* homomorphism (準同型写像) *if*

$$\varphi(vw) = \varphi(v)\varphi(w) \quad \text{for all } v, w \in \Sigma^* .$$

Furthermore, φ is called a λ-free homomorphism (λ-自由な準同型写像), if additionally

$$h(w) = \lambda \quad \text{implies} \quad w = \lambda \quad \text{for all } w \in \Sigma^* .$$

Moreover, if $\varphi \colon \Sigma^* \longrightarrow \Delta^*$ is a homomorphism then we define the inverse (逆写像) *of the homomorphism φ to be the mapping $\varphi^{-1} \colon \Delta^* \longrightarrow \wp(\Sigma^*)$ by setting $\varphi^{-1}(s) = \{w \mid w \in \Sigma^* \text{ and } \varphi(w) = s\}$ for each $s \in \Delta^*$.*

So, a homomorphism is a substitution that maps every symbols of Σ to a *singleton* set. Clearly, by the definition of homomorphism, it already suffices to declare the mapping φ for the symbols in Σ. Note that, when dealing with homomorphisms, we usually identify the language containing exactly one string by the string itself, i.e., instead of $\{s\}$ we shortly write s.

Example 8.2 Let $\Sigma = \{0, 1\}$ and let $\Delta = \{a, b\}$. Then, the mapping $\varphi \colon \Sigma^* \longrightarrow \Delta^*$ defined by $\varphi(0) = ab$ and $\varphi(1) = \lambda$ is a homomorphism but not a λ-free homomorphism. Applying φ to 1100 yields $\varphi(1100) = \varphi(1)\varphi(1)\varphi(0)\varphi(0) = \lambda\lambda abab = abab$ and to the language $\underline{1\langle 0 \rangle 1}$ gives $\varphi(\underline{1\langle 0 \rangle 1}) = \langle \underline{ab} \rangle$.

For seeing the importance of the notions just introduced, consider the language $L = \{a^n b^n \mid n \in \mathbb{N}\}$. This language is context-free as we have shown (cf. Theorem 6.1). Thus, we intuitively know that $\{0^n 1^n \mid n \in \mathbb{N}\}$ is also context-free, because we could go through the grammar and replace all occurrences of a by 0 and all occurrences of b by 1. This observation would suggest that if we replace all occurrences of a and b by strings v and w, respectively, we also get a context-free language. However, it is much less intuitive that we also obtain a context-free language if all occurrences of a and b are replaced by context-free sets of strings V and W, respectively. Nevertheless, we just aim to prove this *closure* property. For the sake of representation, in the following we always assume two finite alphabets Σ and Δ as in Definition 8.1.

Theorem 8.1 $\mathcal{CF}$ *is closed under substitutions.*

Proof. Let $L \in \mathcal{CF}$ be arbitrarily fixed and let τ be a substitution such that $\tau(a)$ is a context-free language for all $a \in \Sigma$. We have to show that $\tau(L)$ is context-free. We shall do this by providing a context-free grammar $\overline{\mathcal{G}} = [\overline{T}, \overline{N}, \overline{\sigma}, \overline{P}]$ such that $L(\overline{\mathcal{G}}) = \tau(L)$.

Since $L \in \mathcal{CF}$, there exists a context-free grammar $\mathcal{G} = [\Sigma, N, \sigma, P]$ in CNF such that $L = L(\mathcal{G})$. Next, let $\Sigma = \{a_1, \ldots, a_n\}$ and consider $\tau(a)$

8.1 Substitutions and Homomorphisms

for all $a \in \Sigma$. By assumption, $\tau(a) \in \mathcal{CF}$ for all $a \in \Sigma$. Thus, there are context-free grammars $\mathcal{G}_a = [T_a, N_a, \sigma_a, P_a]$ such that $\tau(a) = L(\mathcal{G}_a)$ for all $a \in \Sigma$. Without loss of generality, we can assume the sets $N, N_{a_1}, \ldots, N_{a_n}$ to be pairwise disjoint and disjoint to all terminal alphabets considered.

At this point we need an idea how to proceed. For getting an idea, we look at possible derivations in $\mathcal{G}$. Suppose we have a derivation

$$\sigma \stackrel{*}{\underset{\mathcal{G}}{\Longrightarrow}} x_1 x_2 \cdots x_m ,$$

where all $x_i \in \Sigma$ for $i = 1, \ldots, m$. Then, since $\mathcal{G}$ is in CNF, we can conclude that there must be productions $(h_{x_i} \to x_i) \in P$, $i = 1, \ldots, m$, and hence, we easily achieve the following

$$\sigma \stackrel{*}{\underset{\mathcal{G}}{\Longrightarrow}} h_{x_1} h_{x_2} \cdots h_{x_m} \stackrel{m}{\underset{\mathcal{G}}{\Longrightarrow}} x_1 x_2 \cdots x_m , \qquad (8.1)$$

where all $h_{x_i} \in N$. Taking into account that the image $\tau(x_1 \cdots x_m)$ is obtained by calculating $\tau(x_1)\tau(x_2) \cdots \tau(x_m)$, we see that for every string $w_1 w_2 \cdots w_m$ in this image there must be a derivation

$$\sigma_{x_i} \stackrel{*}{\underset{\mathcal{G}_{x_i}}{\Longrightarrow}} w_i , \qquad i = 1, \ldots, m .$$

This directly yields the idea for constructing $\overline{\mathcal{G}}$. That is, we aim to cut the derivation in (8.1) when having obtained $h_{x_1} h_{x_2} \cdots h_{x_m}$. Then, instead of deriving $x_1 x_2 \cdots x_m$, all we need is to generate $\sigma_{x_1} \cdots \sigma_{x_m}$, and thus, we have to replace the productions $(h_{x_i} \to x_i) \in P$ by $(h_{x_i} \to \sigma_{x_i}) \in \overline{P}$, $i = 1, \ldots, m$.

Formalizing this idea yields the following definition of $\overline{\mathcal{G}} = [\overline{T}, \overline{N}, \overline{\sigma}, \overline{P}]$, where

$$\overline{T} = \bigcup_{a \in \Sigma} T_a ,$$

$$\overline{N} = N \cup \left(\bigcup_{a \in \Sigma} N_a\right),$$

$$\overline{\sigma} = \sigma,$$

$$\overline{P} = \left(\bigcup_{a \in \Sigma} P_a\right) \cup P[a /\!/ \sigma_a].$$

Note that $P[a /\!/ \sigma_a]$ is the set of all productions from P, where those productions containing an $a \in \Sigma$ on the right hand side, i.e., $h_a \to a$, are replaced by $h_a \to \sigma_a$.

It remains to show that $\tau(L) = L(\overline{\mathcal{G}})$.

Claim 1. $\tau(L) \subseteq L(\overline{\mathcal{G}})$.

If $\sigma \underset{\mathcal{G}}{\overset{*}{\Longrightarrow}} x_1 \cdots x_m$, where $x_i \in \Sigma$ for all $i = 1, \ldots, m$ and if $\sigma_{x_i} \underset{\mathcal{G}_{x_i}}{\overset{*}{\Longrightarrow}} w_i$, where $w_i \in T_{x_i}^*$, $i = 1, \ldots, m$, then we can also derive $x_1 \cdots x_m$ in the following way:

$$\sigma \underset{\mathcal{G}}{\overset{*}{\Longrightarrow}} h_{x_1} \cdots h_{x_m} \underset{\mathcal{G}}{\overset{*}{\Longrightarrow}} x_1 \cdots x_m,$$

where all $h_{x_i} \in N$. By construction, we can thus generate

$$\sigma \underset{\overline{\mathcal{G}}}{\overset{*}{\Longrightarrow}} h_{x_1} \cdots h_{x_m} \underset{\overline{\mathcal{G}}}{\overset{*}{\Longrightarrow}} \sigma_{x_1} \cdots \sigma_{x_m} \underset{\overline{\mathcal{G}}}{\overset{*}{\Longrightarrow}} w_1 \cdots w_m.$$

Hence, Claim 1 follows.

Claim 2. $L(\overline{\mathcal{G}}) \subseteq \tau(L)$.

Now, we start from $\sigma \overset{*}{\Rightarrow} w$, where $w \in \overline{T}^*$. If $w = \lambda$, then also $\sigma \to \lambda$ is in P, and we are done. Otherwise, the construction of $\overline{\mathcal{G}}$ ensures that the derivation of w must look as follows

$$\sigma \underset{\overline{\mathcal{G}}}{\overset{*}{\Longrightarrow}} \sigma_{x_1} \cdots \sigma_{x_m} \underset{\overline{\mathcal{G}}}{\overset{*}{\Longrightarrow}} w.$$

By construction, we then know that $\sigma \underset{\mathcal{G}}{\overset{*}{\Longrightarrow}} x_1 \cdots x_m$, see (8.1). Additionally, there are strings $w_1, \ldots, w_m \in \overline{T}^*$ such that $w = w_1 \cdots w_m$ and $\sigma_{x_i} \underset{\mathcal{G}_{x_i}}{\overset{*}{\Longrightarrow}} w_i$ for all $i = 1, \ldots, m$. Consequently, $w_i \in \tau(x_i)$. Therefore, $w \in \tau(L)$ and we are done.

Putting it all together, we see that $\tau(L) = L(\overline{\mathcal{G}})$. ∎

8.1 Substitutions and Homomorphisms

Please note that we have skipped some formal parts in the above proof to make it easier to read and to understand. However, we should be aware of this omission. Since we omit such tedious formal parts in future proofs, this is a good place to tell what one should do when reading this book.

The omitted part is formally verified as follows. We define a mapping

$$\zeta \colon (\Sigma \cup N) \longrightarrow N \cup \left(\bigcup_{a \in \Sigma} \{\sigma_a\} \right)$$

by setting

$$\zeta(a) = \begin{cases} a, & \text{if } a \in N\,; \\ \sigma_a, & \text{if } a \in \Sigma\,. \end{cases}$$

We extend ζ to strings by defining

$\zeta(\lambda) = \lambda$, and

$\zeta(x) = \zeta(x_1)\zeta(x_2)\cdots\zeta(x_m) \quad$ for all $x \in (\Sigma \cup N)^*$, $x = x_1 \cdots x_m$.

Furthermore, we define a grammar $\mathcal{G}' = [T', N', \sigma, P']$, where

$$T' = \bigcup_{a \in \Sigma} \{\sigma_a\},$$
$$N' = N,$$
$$P' = \{A \rightarrow \zeta(\alpha) \mid (A \rightarrow \alpha) \in P\}.$$

Then we can prove the following.

Claim 1. For all $A \in N$ we have $A \underset{\mathcal{G}}{\overset{*}{\Longrightarrow}} \alpha$ if and only if $A \underset{\mathcal{G}'}{\overset{*}{\Longrightarrow}} \zeta(\alpha)$.

Claim 1 is intuitively obvious, since ζ corresponds to a simple renaming of Σ. The formal proof is done by induction on the length of the derivation and left as an exercise.

Using Claim 1, in particular, we have the following

$$\sigma \underset{\mathcal{G}}{\overset{*}{\Longrightarrow}} x_1 \cdots x_m \in L(\mathcal{G}) \quad \text{if and only if} \quad \sigma \underset{\mathcal{G}'}{\overset{*}{\Longrightarrow}} \sigma_{x_1} \cdots \sigma_{x_m},$$

where $x_i \in \Sigma$ for all $i = 1, \ldots, m$. This completes the formal verification.

Theorem 8.1 allows for the following nice corollary.

Corollary 8.1 $\mathcal{CF}$ *is closed under homomorphisms.*

Proof. Since homomorphisms are a special type of substitution, it suffices to argue that every singleton subset is context-free. But this is obvious, because we have already shown that every finite language belongs to $\mathcal{REG}$ and that $\mathcal{REG} \subseteq \mathcal{CF}$. Thus, the corollary follows. ∎

Now, we encourage the reader to solve the following exercises.

Exercise 8.1 *Let* $\Sigma = \{0, 1, 2\}$, $\Delta = \{a, b\}$, $L = \{0^n 1 2^n \mid n \in \mathbb{N}^+\}$, *and let the substitution* $\tau \colon \Sigma \longrightarrow \wp(\Delta^*)$ *be defined as follows:*
$\tau(0) = L_{pal}$, $\tau(1) = \{a^{3i} b^i \mid i \in \mathbb{N}^+\}$, *and* $\tau(2) = \{a, b\}^+$.
Compute $\tau(L)$ *and provide a grammar for* $\tau(L)$.

Exercise 8.2 *Prove or disprove:* $\mathcal{REG}$ *is closed under substitutions.*

Recall that the family of all languages generated by a grammar in the sense of Definition 2.1 is denoted by $\mathcal{L}_0$.

Exercise 8.3 *Figure out why the proof of Theorem 8.1 does not work for showing that* $\mathcal{L}_0$ *is closed under substitutions. Then provide the necessary modifications to show closure under substitution for the family* $\mathcal{L}_0$.

8.2 Homomorphic Characterization of $\mathcal{CF}$

When we started to study context-free languages, we emphasized that many programming languages use balanced brackets of different kinds. Therefore, we continue with a closer look at bracket languages. Such languages are called *Dyck languages*[†] (ダイク言語).

In order to define Dyck languages, we need the following notations. Let

---

[†] Walter von Dyck (1856-1934) was a mathematician. He was a founder of combinatorial group theory.

8.2 Homomorphic Characterization of $\mathcal{CF}$

$n \in \mathbb{N}^+$ and let $X_n = \{a_1, \overline{a}_1, a_2, \overline{a}_2, \ldots, a_n, \overline{a}_n\}$. We consider the set X_n as a set of different bracket symbols, where a_i is an opening bracket and $\overline{a}_i$ is the corresponding closing bracket. Thus, it is justified to speak of X_n as a set of n different bracket symbols.

Definition 8.4 *A language* L *is said to be a* Dyck *language with* n *bracket symbols if* L *is isomorphic (* 同型 *) to the language* D_n *generated by the following grammar* $\mathcal{G}_n = [X_n, \{\sigma\}, \sigma, P_n]$, *where* P_n *is given by*

$$P_n = \{\sigma \to \lambda,\ \sigma \to \sigma\sigma,\ \sigma \to a_1\sigma\overline{a}_1,\ \ldots,\ \sigma \to a_n\sigma\overline{a}_n\}.$$

The importance of Dyck languages will become immediately transparent, since we are going to prove a beautiful characterization theorem for context-free languages by using them, i.e., the Chomsky-Schützenberger theorem.

Theorem 8.2 (Chomsky-Schützenberger Theorem [4][†]**)** *For every context-free language* L *there is an* $n \in \mathbb{N}^+$, *a homomorphism* h *and a regular language* R_L *such that* $L = h(D_n \cap R_L)$.

Proof. Consider any language $L \in \mathcal{CF}$. Without loss of generality we can assume that $\lambda \notin L$. Let $\mathcal{G} = [T, N, \sigma, P]$ be a context-free grammar in CNF such that $L = L(\mathcal{G})$. Let $T = \{x_1, \ldots, x_m\}$ and consider all productions in P. Since $\mathcal{G}$ is in CNF, all productions have the form $h_i \to h_i' h_i''$ or $h_j \to x$. Let t be the number of all nonterminal productions, i.e., of all productions $h_i \to h_i' h_i''$. Note that for any two such productions it is well possible that some but not all nonterminal symbols coincide.

In all we have m terminal symbols and t nonterminal productions. Thus, we try the Dyck language D_{m+t} over

$$X_{m+t} = \{\overline{x}_1, \ldots, \overline{x}_m, \overline{x}_{m+1}, \ldots, \overline{x}_{m+t}, x_{m+1}, \ldots, x_{m+t}, x_1, \ldots, x_m\}.$$

Next, we consider the mapping $\chi_{m+t}: X_{m+t} \longrightarrow T^*$ defined as follows

---

[†] チョムスキー-シュツェンベルガーの定理

8. CF and Homomorphisms

$$\chi_{m+t}(x_j) = \begin{cases} x_j, & \text{if } 1 \leq j \leq m \ ; \\ \lambda, & \text{if } m+1 \leq j \leq m+t \ ; \end{cases}$$

and $\chi_{m+t}(\bar{x}_j) = \lambda$ for all $j = 1, \ldots, m+t$. We leave it as an exercise to show that χ_{m+t} is a homomorphism.

Now we define the following grammar $\mathcal{G}_L = [X_{m+t}, N, \sigma, P_L]$, where

$$P_L = \{h \to x_i \bar{x}_i \mid 1 \leq i \leq m \text{ and } (h \to x_i) \in P\}$$

$$\cup \ \{h \to x_i \bar{x}_i \bar{x}_{m+j} h''_j \mid 1 \leq i \leq m, \ (h \to x_i) \in P, \ 1 \leq j \leq t\}$$

$$\cup \ \{h_j \to x_{m+j} h'_j \mid 1 \leq j \leq t\} \ .$$

Clearly, $\mathcal{G}_L$ is a regular grammar. We set $R_L = L(\mathcal{G}_L)$, and prove that

$$L = \chi_{m+t}(D_{m+t} \cap R_L) \ .$$

This is done via the following claims and lemmata.

Claim 1. $L \subseteq \chi_{m+t}(D_{m+t} \cap R_L)$.

The proof of Claim 1 is mainly based on the following lemma.

Lemma 8.1 *Let $\mathcal{G}$ be the grammar for L fixed above, let $\mathcal{G}_L$ be the grammar for R_L and let $h \in N$. If*

$$h \underset{\mathcal{G}}{\overset{1}{\Longrightarrow}} w_1 \underset{\mathcal{G}}{\overset{1}{\Longrightarrow}} w_2 \underset{\mathcal{G}}{\overset{1}{\Longrightarrow}} \cdots \underset{\mathcal{G}}{\overset{1}{\Longrightarrow}} w_{n-1} \underset{\mathcal{G}}{\overset{1}{\Longrightarrow}} w_n \in T^*$$

then there exists a $q \in D_{m+t}$ such that $h \underset{\mathcal{G}_L}{\overset{*}{\Longrightarrow}} q$ and $\chi_{m+t}(q) = w_n$.

Proof. The lemma is shown by induction on the length n of the derivation. For the induction basis let $n = 1$. Thus, our assumption is that

$$h \underset{\mathcal{G}}{\overset{1}{\Longrightarrow}} w_1 \in T^* \ .$$

Since $\mathcal{G}$ is in CNF, we can conclude that $(h \to w_1) \in P$. So, by the definition of CNF, we must have $w_1 = x$ for some $x \in T$.

8.2 Homomorphic Characterization of $\mathcal{CF}$

We have to show that there is a $q \in D_{m+t}$ such that $h \underset{g_L}{\overset{*}{\Longrightarrow}} q$ and $\chi_{m+t}(q) = x$. By construction, the production $h \to x\bar{x}$ belongs to P_L (cf. the first set of the definition of P_L). Thus, we can simply set $q = x\bar{x}$. Now, the induction basis follows, since the definition of χ_{m+t} directly yields

$$\chi_{m+t}(q) = \chi_{m+t}(x\bar{x}) = \chi_{m+t}(x)\chi_{m+t}(\bar{x}) = x\lambda = x.$$

Assuming the induction hypothesis for $n \geq 1$, we are going to perform the induction step to $n+1$. So, let

$$h \underset{g}{\overset{1}{\Longrightarrow}} w_1 \underset{g}{\overset{1}{\Longrightarrow}} \cdots \underset{g}{\overset{1}{\Longrightarrow}} w_n \underset{g}{\overset{1}{\Longrightarrow}} w_{n+1} \in T^*$$

be a derivation of length $n+1$. Because of $n \geq 1$, and since the derivation has length at least 2, we can conclude that the production used to derive w_1 must be of the form $h \to h'h''$, where $h, h', h'' \in N$. Therefore, there must be a j such that $1 \leq j \leq t$ and $h = h_j$ as well as $w_1 = h'_j h''_j$.

Consequently, there must be v_1, v_2 such that $w_{n+1} = v_1 v_2$ and

$$h'_j \underset{g}{\overset{*}{\Longrightarrow}} v_1 \quad \text{and} \quad h''_j \underset{g}{\overset{*}{\Longrightarrow}} v_2.$$

Since the length of the complete derivation is $n+1$, both the generation of v_1 and of v_2 must have a length smaller than or equal to n.

Hence, we can apply the induction hypothesis. That is, there are strings q_1 and q_2 such that $q_1, q_2 \in D_{m+t}$ and $\chi_{m+t}(q_1) = v_1$ as well as $\chi_{m+t}(q_2) = v_2$. Furthermore, by the induction hypothesis we know that

$$h'_j \underset{g_L}{\overset{*}{\Longrightarrow}} q_1 \quad \text{and} \quad h''_j \underset{g_L}{\overset{*}{\Longrightarrow}} q_2.$$

Taking into account that $(h_j \to h'_j h''_j) \in P$, we know by construction that $h_j \to x_{m+j} h'_j$ is a production in P_L. Thus,

$$h = h_j \underset{g_L}{\overset{1}{\Longrightarrow}} x_{m+j} h'_j \underset{g_L}{\overset{*}{\Longrightarrow}} x_{m+j} q_1$$

8. CF and Homomorphisms

is a regular derivation. Moreover, the last step of this derivation must look as follows

$$x_{m+j} q_1' h_k \underset{\mathcal{G}_L}{\overset{1}{\Longrightarrow}} x_{m+j} q_1' x \bar{x} .$$

where $h_k \to x\bar{x}$ is the rule applied and where x is determined by the condition $q_1 = q_1' x\bar{x}$.

Now, we replace this step by using the production $h_k \to x \overline{x} x_{m+j} h_j''$ which also belongs to P_L. Thus, we obtain

$$h = h_j \underset{\mathcal{G}_L}{\overset{1}{\Longrightarrow}} x_{m+j} h_j' \underset{\mathcal{G}_L}{\overset{*}{\Longrightarrow}} x_{m+j} q_1 \bar{x}_{m+j} h_j'' \underset{\mathcal{G}_L}{\overset{*}{\Longrightarrow}} x_{m+j} q_1 \bar{x}_{m+j} q_2 =: q \in D_{m+t}.$$

The containment in D_{m+t} is due to the correct usage of the brackets x_{m+j} and $\bar{x}_{m+j}$ around q_1 and the fact that $q_2 \in D_{m+t}$ as well as by the definition of the Dyck language. Finally, the definition of χ_{m+t} ensures that $\chi_{m+t}(x_{m+j} q_1 \bar{x}_{m+j} q_2) = v_1 v_2$. This proves Lemma 8.1.

Now, Claim 1 immediately follows for $h = \sigma$.

Claim 2. $L \supseteq \chi_{m+t}(D_{m+t} \cap R_L)$.

The proof of Claim 2 is mainly based on the following lemma.

Lemma 8.2 *Let $\mathcal{G}$ be the grammar for L fixed above, let $\mathcal{G}_L$ be the grammar for R_L and let $h \in N$. If*

$$h \underset{\mathcal{G}_L}{\overset{1}{\Longrightarrow}} w_1 \underset{\mathcal{G}_L}{\overset{1}{\Longrightarrow}} \cdots \underset{\mathcal{G}_L}{\overset{1}{\Longrightarrow}} w_n \in D_{m+t}$$

then $h \underset{\mathcal{G}}{\overset{*}{\Longrightarrow}} \chi_{m+t}(w_n)$.

Proof. Again, the lemma is shown by induction on the length of the derivation. We perform the induction basis for $n = 1$. Consider

$$h \underset{\mathcal{G}_L}{\overset{1}{\Longrightarrow}} w_1 \in D_{m+t} .$$

Hence, we must conclude that $(h \to w_1) \in P_L$. So, there must exist $x_i \bar{x}_i$ such that $w_1 = x_i \bar{x}_i$, $1 \leq i \leq m$ and $(h \to x_i x_i) \in P_L$. By the definition of P_L we conclude that $(h \to x_i) \in P$. Hence

$$h \underset{g}{\overset{1}{\Longrightarrow}} x_i = \chi_{m+t}(x_i\bar{x}_i) = \chi_{m+t}(w_1) \ .$$

This proves the induction basis.

Next, assume the induction hypothesis for all derivation of length less than or equal to n. We perform the induction step from n to $n+1$. Consider

$$h \underset{g_L}{\overset{1}{\Longrightarrow}} w_1 \underset{g_L}{\overset{1}{\Longrightarrow}} \cdots \underset{g_L}{\overset{1}{\Longrightarrow}} w_n \underset{g_L}{\overset{1}{\Longrightarrow}} w_{n+1} \in D_{m+t} \ .$$

The derivation $h \underset{g_L}{\overset{1}{\Longrightarrow}} w_1$ must have been done by using a production $(h \to w_1) \in P_L$, where

$$w_1 \in \{x\bar{x}\chi_{m+j}h_j'',\ \chi_{m+j}h_j'\} \ .$$

But $w_1 \neq x\bar{x}\chi_{m+j}h_j''$, since $x\bar{x}\chi_{m+j}$ cannot be removed by any further derivation step. This would imply $w_{n+1} = x\bar{x}\chi_{m+j}r \notin D_{m+t}$. So, this case cannot happen.

Thus, the only remaining case is that $w_1 = \chi_{m+j}h_j'$. Hence there is a j with $1 \leq j \leq t$ such that $h = h_j$ and $w_1 = \chi_{m+j}h_j'$. This implies that there must be a derivation

$$h_j' \underset{g_L}{\overset{1}{\Longrightarrow}} w_1' \underset{g_L}{\overset{1}{\Longrightarrow}} w_2' \underset{g_L}{\overset{1}{\Longrightarrow}} \cdots \underset{g_L}{\overset{1}{\Longrightarrow}} w_{n+1}' \ ,$$

where $w_{n+1} = \chi_{m+j}w_{n+1}' \in D_{m+t}$. Therefore, there are w' and w'' such that

$$w_{n+1} = \chi_{m+j}w'\bar{x}_{m+j}w'' \quad \text{and} \quad w',w'' \in D_{m+t} \ .$$

Hence, there exists a k with $2 \leq k \leq m$ such that $w_k' = w'\bar{x}_{m+j}h_j''$. This means nothing else than having used in the kth derivation step a production of the form $h \to x\bar{x}\chi_{m+j}h_j''$. Consequently, $(h \to x) \in P$ and $(h \to x\bar{x}) \in P_L$.

88   8. CF and Homomorphisms

We thus replace the application of $h \to x\overline{xx}_{m+j}h_j''$ by an application of $h \to x\overline{x}$ and obtain

$$h_j \underset{\mathcal{G}_L}{\overset{*}{\Longrightarrow}} w' \quad \text{and} \quad h_j' \underset{\mathcal{G}_L}{\overset{*}{\Longrightarrow}} w'',$$

where both derivations have a length less than or equal to n.

Applying the induction hypothesis yields

$$h_j' \underset{\mathcal{G}}{\overset{*}{\Longrightarrow}} \chi_{m+t}(w')$$

$$h_j'' \underset{\mathcal{G}}{\overset{*}{\Longrightarrow}} \chi_{m+t}(w'') \quad \text{and thus}$$

$$h \underset{\mathcal{G}}{\overset{1}{\Longrightarrow}} h_j'h_j'' \underset{\mathcal{G}}{\overset{*}{\Longrightarrow}} \chi_{m+t}(w'w'') = \chi_{m+t}(x_{m+j}w'\overline{x}_{m+j}w'')$$

$$= \chi_{m+t}(w_{n+1}).$$

This proves Lemma 8.2. Furthermore, Claim 2 is a direct consequence of Lemma 8.2 for $h = \sigma$.

Claim 1 and Claim 2 together imply the theorem.   ∎

Note that the Chomsky-Schützenberger Theorem has a nice counterpart which can be stated in terms of languages accepted by pushdown automata.

Finally, we present an example illustrating the construction in the proof of the Chomsky-Schützenberger Theorem.

Example 8.3 *Let the context-free grammar* $\mathcal{G} = [\{a, b\}, \{\sigma, h\}, \sigma, P]$ *be given, where* $P = \{\sigma \to hbh,\ h \to hah,\ h \to ab\}$, *and consider* $L = L(\mathcal{G})$.

We perform the following steps. First, we construct a separated grammar $\mathcal{G}_{sep}$ such that $L(\mathcal{G}_{sep}) = L(\mathcal{G})$. Then, we transform $\mathcal{G}_{sep}$ into CNF and obtain $\mathcal{G}_{CNF}$ such that $L = L(\mathcal{G}_{CNF})$. Finally, we determine n, the homomorphism χ, the Dyck language D_n and the regular language R_L such that $L = \chi(D_n \cap R_L)$.

Note that $\mathcal{G}$ is already reduced but not separated.

8.2 Homomorphic Characterization of $\mathcal{CF}$

Step 1. Construction of $\mathcal{G}_{sep}$.

We apply the algorithm given in the proof of Theorem 7.1 and introduce the new nonterminals h_a, h_b. Now, we obtain the separated grammar $\mathcal{G}_{sep} = [\{a, b\}, \{\sigma, h, h_a, h_b\}, \sigma, P_{sep}]$, where

$$P_{sep} = \{\sigma \rightarrow hh_bh,\ h \rightarrow hh_ah,\ h \rightarrow h_ah_b,\ h_a \rightarrow a,\ h_b \rightarrow b\}.$$

Step 2. Construction of $\mathcal{G}_{CNF}$.

Applying the algorithm given in the proof of Theorem 7.2, we skip the first part, since there is no production of the form $h \rightarrow h'$ in P_{sep}. The only productions not yet in CNF are $\sigma \rightarrow hh_bh$ and $h \rightarrow hh_ah$. We introduce the new nonterminals h_{h_ah}, h_{h_bh} and obtain

$$\mathcal{G}_{CNF} = [\{a, b\}, \{\sigma, h, h_a, h_b, h_{h_ah}, h_{h_bh}\}, \sigma, P_{CNF}],\quad \text{where}$$

$$P_{CNF} = \{\sigma \rightarrow hh_{h_bh},\ h_{h_bh} \rightarrow h_bh,\ h \rightarrow hh_{h_ah},\ h_{h_ah} \rightarrow h_ah,$$
$$h \rightarrow h_ah_b,\ h_a \rightarrow a,\ h_b \rightarrow b\}.$$

Step 3. Performing the construction of the proof of Theorem 8.2.

Clearly, $m = \text{card}(\{a, b\}) = 2$ and $t = 5$, since we have five productions of the form $h \rightarrow h'h''$. Thus we set

$$X_{2+5} = \{\bar{a}, \bar{b}, \bar{x}_3, \bar{x}_4, \bar{x}_5, \bar{x}_6, \bar{x}_7, x_3, x_4, x_5, x_6, x_7, a, b\},$$

where we used a, b instead of x_1 and x_2 and also $\bar{a}$, $\bar{b}$ instead of $\bar{x}_1$, $\bar{x}_2$.

The homomorphism χ_{2+5} is defined as follows

$\chi_{2+5}(a) = a$,

$\chi_{2+5}(b) = b$ and

$\chi_{2+5}(y) = \lambda$ for all $y \in X_{2+5} \setminus \{a, b\}$.

So we have $n = 7$ and the Dyck language D_7 with the bracket symbols from X_{2+5}.

8. $\mathcal{CF}$ and Homomorphisms

Next, we have to construct the grammar $\mathcal{G}_L = [X_{2+5}, N, \sigma, P_L]$ for the regular language $R_L = L(\mathcal{G}_L)$. Here we obtain

$N = \{\sigma, h, h_a, h_b, h_{h_a h}, h_{h_b h}\}$ the nonterminals from $\mathcal{G}_{CNF}$

$P_L = P_L^I \cup P_L^{II} \cup P_L^{III}$ corresponding to the 3 sets

$P_L^I = \{h_a \rightarrow a\bar{a}, \; h_b \rightarrow b\bar{b}\}$,

$P_L^{II} = \{h_a \rightarrow a\bar{a}x_3 h_{h_b h}, \; h_a \rightarrow a\bar{a}x_4 h, \; h_a \rightarrow a\bar{a}x_5 h_{h_a h},$

$\quad h_a \rightarrow a\bar{a}x_6 h, \; h_a \rightarrow a\bar{a}x_7 h_b,$

$\quad h_b \rightarrow b\bar{b}\bar{x}_3 h_{h_b h}, \; h_b \rightarrow b\bar{b}\bar{x}_4 h, \; h_b \rightarrow b\bar{b}\bar{x}_5 h_{h_a h},$

$\quad h_b \rightarrow b\bar{b}\bar{x}_6 h, \; h_b \rightarrow a\bar{b}\bar{x}_7 h_b\}$, and

$P_L^{III} = \{\sigma \rightarrow x_3 h, \; h_{h_b h} \rightarrow x_4 h_b, \; h \rightarrow x_5 h, \; h_{h_a h} \rightarrow x_6 h_a,$

$\quad h \rightarrow x_7 h_a\}$.

Note that the five productions in P_L^{II} having $a\bar{a}$ in their right hand side and the five productions in P_L^{II} having $b\bar{b}$ in their right hand side as well as the five productions in P_L^{III} correspond to the following five productions from P_{CNF}

$\sigma \rightarrow h h_{h_b h}, \; h_{h_b h} \rightarrow h_b h, \; h \rightarrow h h_{h_a h}, \; h_{h_a h} \rightarrow h_a h, \; h \rightarrow h_a h_b$,

which have the form $h \rightarrow h' h''$.

Furthermore, we present two derivations. First, we derive $abbab$ in $\mathcal{G}_{CNF}$ by using the first, fifth, second and fifth production and then productions six and seven until all nonterminals are replaced:

$\sigma \Rightarrow h h_{h_b h} \Rightarrow h_a h_b h_{h_b h} \Rightarrow h_a h_b h_b h \Rightarrow h_a h_b h_b h_a h_b \overset{*}{\Rightarrow} abbab$.

Second, we provide a derivation for $q \in D_7$ by using $\mathcal{G}_L$ such that $\chi_{2+5}(q) = abbab$.

$\sigma \Rightarrow x_3 h \Rightarrow x_3 x_7 h_a \Rightarrow x_3 x_7 a \overline{ax}_7 h_b \Rightarrow x_3 x_7 a \overline{ax}_7 b \overline{bx}_3 h_{h_b h}$

$\Rightarrow x_3 x_7 a \overline{ax}_7 b \overline{bx}_3 x_4 h_b \Rightarrow x_3 x_7 a \overline{ax}_7 b \overline{bx}_3 x_4 b \overline{bx}_4 h$

$\Rightarrow x_3 x_7 a \overline{ax}_7 b \overline{bx}_3 x_4 b \overline{bx}_4 x_7 h_a \Rightarrow x_3 x_7 a \overline{ax}_7 b \overline{bx}_3 x_4 b \overline{bx}_4 x_7 a \overline{ax}_7 h_b$

$\Rightarrow x_3 x_7 a \overline{ax}_7 b \overline{bx}_3 x_4 b \overline{bx}_4 x_7 a \overline{ax}_7 b \overline{b} =: q$.

Clearly, $\chi_{2+5}(x_3 x_7 a \overline{ax}_7 b \overline{bx}_3 x_4 b \overline{bx}_4 x_7 a \overline{ax}_7 b \overline{b}) = abbab$. We leave it as an exercise to derive q by using the grammar for D_7.

Now, the reader is encouraged to solve the following exercise.

Exercise 8.4 *Consider the following language* $L = D_1 \cup L_{pal2}$.

(1) *Prove that L is context-free.*

(2) *Determine a number* $n \in \mathbb{N}^+$, *a homomorphism* h *and a regular language* R_L *such that* $L = h(D_n \cap R_L)$. *Prove your assertion.*

8.3 Problem Set 8

Problem 8.1 Let $\varphi: \{a, b\}^* \longrightarrow \{a, b\}^*$ be the homomorphism defined by $\varphi(a) = b$ and $\varphi(b) = ab$. Furthermore, we set $\varphi^1(w) = \varphi(w)$ and $\varphi^{n+1}(w) = \varphi(\varphi^n(w))$ for every $n \in \mathbb{N}$, $n \geq 1$, and $w \in \{a, b\}^*$. Determine the length of the string $\varphi^n(a)$ in dependence on n, where $n \in \mathbb{N}^+$.

Problem 8.2 Prove or disprove: $\mathcal{REG}$ is closed under homomorphisms.

Problem 8.3 Prove or disprove: $\mathcal{REG}$ is closed under inverse homomorphisms.

9 Pushdown Automata

9.1 Introducing Pushdown Automata

As already mentioned, the context-free languages also have a type of automaton that characterizes them. This automaton, called *pushdown automaton* (プッシュダウンオートマトン) , is an extension of the NFA with λ-transitions. By λ-transition we mean that the automaton is allowed to read the empty word λ on its input tape and to change its state accordingly.

A pushdown automaton (abbr. PDA) is essentially an NFA with λ-transitions with the addition of a stack (スタック). The stack allows a PDA to memorize any finite string, but the access to the stack is in lifo-mode only (lifo stands for "last in first out"). The stack can be *read* (リード), *pushed* (プッシュ), and *popped* (ポップ) only at the top (cf. Figure 9.1).

The PDA shown in Figure 9.1 informally works as follows. A finite state control reads inputs, one symbol at a time or it reads λ instead of an input symbol. Moreover, it reads the symbol at the top of the stack. The PDA bases its state transition on its current state, the input symbol (or λ), and the symbol at the top of its stack. If λ is read instead of the input symbol, then we say that the PDA makes a *spontaneous transition* (自発的遷移). In one transition, the PDA:

9.1 Introducing Pushdown Automata

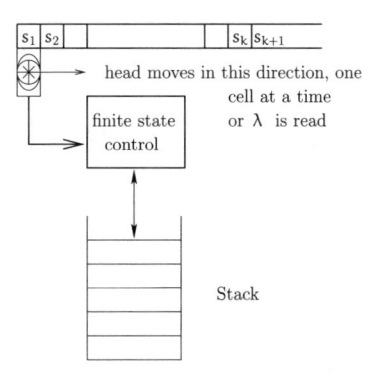

Figure 9.1 A PDA.

(1) Consumes from the input the symbol it reads. If λ is read then no input symbol is consumed.

(2) Goes to a new state (which may or may not be the same state as its current state).

(3) Replaces the symbol at the top of the stack by any string. The string could be λ, which corresponds to a pop of the stack. It could be the same symbol that appeared at the top of the stack previously, i.e., no change is made to the stack. It could also replace the symbol on top of the stack by one other symbol. In this case, the PDA changes the top of the stack but does neither push or pop it.

Finally, the top stack symbol could be replaced by two or more symbols which has the effect of (possibly) changing the top stack symbol and then pushing one or more new symbols onto the stack.

Example 9.1 We informally show $L_{pal2} = \{ww^T \mid w \in \{0,1\}^*\}$ to be acceptable by a PDA. Let $x \in \{0,1\}^*$ be given as input.

(1) Start in a state q_0 representing a "guess" that we have not yet seen the middle of x. While in state q_0, we read one symbol at a time and store the symbol read in the stack by pushing a copy of each input symbol onto the stack.

(2) At any time, we may guess that we have seen the middle (i.e., the end of w if $x = ww^T$ is an input string from L). At this time, w will be on the stack with the rightmost symbol of w at the top and the leftmost symbol of w at the bottom. We signify this choice by spontaneously changing the state to q_1 (i.e., we read λ instead of the next input symbol).

(3) Once in state q_1, we compare the input symbols with the symbols at the top of the stack. If the symbol read from input is equal to the symbol at the top of the stack, we proceed in state q_1 and pop the stack. If they are different, we finish without accepting the input. That is, this branch of computation dies.

(4) If we reach the end of x and the stack is empty, then we accept x.

Clearly, if $x \in L$, then by guessing the middle of x rightly, we arrive at an accepting computation path. If $x \notin L$, then independently of what we are guessing, no computation path will lead to acceptance. Thus, the PDA described above is a nondeterministic acceptor for L. ∎

Note that a PDA is allowed to change the stack as described above while performing a spontaneous transition. Before presenting the formal definition of a PDA, this is a good place to think of ways to define the language accepted by a PDA. Looking at Example 9.1, we see that the PDA has finished its computation with empty stack. Thus, it would be natural to define the language accepted by a PDA to be the set of all strings on which the PDA has a computation that ends with empty stack.

Second, we can adopt the method we have used for finite automata. That is, we choose a subset of the set of all states and declare each state in this subset to be an accepting state. If taking this approach, it would be natural to define the language accepted by a PDA to be the set of all strings for

9.1 Introducing Pushdown Automata

which there is a computation ending in an accepting state.

As we shall show below, both method are equivalent. Therefore, we have to define both modes of acceptance here.

We continue with a formal definition of a PDA.

Definition 9.1 *A 7-tuple* $\mathcal{K} = [Q, \Sigma, \Gamma, \delta, q_0, k_0, F]$ *is called a* PDA *if*

(1) Q *is a finite nonempty set (the set of states (状態集合))*,

(2) Σ *is an alphabet (the input alphabet (入力アルファベット))*,

(3) Γ *is an alphabet (the stack alphabet (スタックアルファベット))*,

(4) $\delta \colon Q \times (\Sigma \cup \{\lambda\}) \times \Gamma \longrightarrow \wp_{fin}(Q \times \Gamma^*)$, *the* transition relation[†] (遷移関係),

(5) $q_0 \in Q$ *is the* initial state (初期状態),

(6) k_0 *is the so-called* stack symbol (スタック記号), *i.e.*, $k_0 \in \Gamma$ *and initially the stack contains exactly one k_0 and nothing else.*

(7) $F \subseteq Q$, *the* set of final states (最終状態の集合).

In the following, unless otherwise stated, we use small letters from the beginning of the alphabet to denote input symbols, and small letters from the end of the alphabet to denote strings of input symbols. Furthermore, we use capital letters to denote stack symbols from Γ and small Greek letters to denote strings of stack symbols.

Next, consider

$$\delta(q, a, Z) = \{(q_1, \gamma_1), (q_2, \gamma_2), \ldots, (q_m, \gamma_m)\},$$

where $q, q_i \in Q$ for $i = 1, \ldots, m$, $a \in \Sigma$, $Z \in \Gamma$ and $\gamma_i \in \Gamma^*$ for $i = 1, \ldots, m$.

The interpretation is that the PDA $\mathcal{K}$ is in state q, reads a on its input tape and Z on the top of its stack. Then it can nondeterministically choose exactly one (q_i, γ_i), $i \in \{1, \ldots, m\}$ for the transition to be made. That is, it changes its internal state to q_i, moves the head on the input tape one

---

[†] Here we use $\wp_{fin}(Q \times \Gamma^*)$ to denote the set of all finite subsets of $Q \times \Gamma^*$.

position to the right provided $a \neq \lambda$ and replaces Z by γ_i. We make the convention that the rightmost symbol of γ_i is pushed first in the stack, then the second symbol (if any) from the right, and so on. So the leftmost symbol of γ_i is the new symbol which is then on the top of the stack. If $\gamma_i = \lambda$, then the interpretation is that Z has been removed from the stack.

If $a = \lambda$, the interpretation is the same as above, except that the head on the input tape is *not* moved.

In order to formally deal with computations performed by a PDA we define *instantaneous descriptions* (時点表示). An instantaneous description is a triple (q, w, γ), where $q \in Q$, $w \in \Sigma^*$ and $\gamma \in \Gamma^*$.

Let $\mathcal{K} = [Q, \Sigma, \Gamma, \delta, q_0, k_0, F]$ be a PDA. Then we write

$$(q, aw, Z\alpha) \xrightarrow[\mathcal{K}]{1} (p, w, \beta\alpha)$$

provided $(p, \beta) \in \delta(q, a, Z)$. Again note that a may be a symbol from Σ or $a = \lambda$. By $\xrightarrow[\mathcal{K}]{*}$ we denote the reflexive-transitive closure of $\xrightarrow[\mathcal{K}]{1}$.

Now we are ready to define the two modes of acceptance.

Definition 9.2 *Let $\mathcal{K} = [Q, \Sigma, \Gamma, \delta, q_0, k_0, F]$ be a PDA. We define the language accepted by $\mathcal{K}$ via final state (最終状態による) to be the set*

$$L(\mathcal{K}) = \{w \mid (q_0, w, k_0) \xrightarrow[\mathcal{K}]{*} (p, \lambda, \gamma) \text{ for some } p \in F \text{ and a } \gamma \in \Gamma^*\}.$$

The language accepted by $\mathcal{K}$ via empty stack (空スタックによる) is the set

$$N(\mathcal{K}) = \{w \mid (q_0, w, k_0) \xrightarrow[\mathcal{K}]{*} (p, \lambda, \lambda) \text{ for some } p \in Q\}.$$

Since the sets of final states is irrelevant if acceptance via empty stack is considered, we always set $F = \emptyset$ in this case.

Exercise 9.1 *Provide a formal definition for a PDA that accepts the language $L_{pal2} = \{ww^T \mid w \in \{0,1\}^*\}$.*

At this point it is only natural to ask what is the appropriate definition of determinism for PDAs. The answer is given by our next definition.

9.1 Introducing Pushdown Automata

Definition 9.3 *A PDA* $\mathcal{K} = [Q, \Sigma, \Gamma, \delta, q_0, k_0, F]$ *is said to be* deterministic *if*

(1) *for every* $q \in Q$ *and* $Z \in \Gamma$ *we have* $\delta(q, a, Z) = \emptyset$ *for all* $a \in \Sigma$ *if* $\delta(q, \lambda, Z) \neq \emptyset$, *and*

(2) *for all* $q \in Q$, $Z \in \Gamma$ *and* $a \in \Sigma \cup \{\lambda\}$ *we have* $\mathrm{card}(\delta(q, a, Z)) \leq 1$.

In Definition 9.3, we had to include Condition (1) to avoid a choice between a normal transition and a spontaneous transition. On the other hand, Condition (2) guarantees that there is no choice in any step. So, Condition (2) resembles the condition we had imposed when defining DFAs. The language accepted by a deterministic PDA is defined in the same way as for nondeterministic PDAs. That is, we again distinguish between acceptance via final state and empty stack, respectively (cf. Problem 9.1).

As far as finite automata have been concerned, we could prove that the class of languages accepted by DFAs is the same as the class of languages accepted by NFAs. Note that an analogous result *cannot be obtained* for PDAs. For example, there is no deterministic PDA accepting the language $L_{pal2} = \{ww^\mathsf{T} \mid w \in \{0,1\}^*\}$.

We continue by comparing the power of the two notions of acceptance. First, we show the following theorem.

Theorem 9.1 *Let* $L = L(\mathcal{K})$ *for a PDA* $\mathcal{K}$. *Then there exists a PDA* $\widetilde{\mathcal{K}}$ *such that* $L = N(\widetilde{\mathcal{K}})$.

Proof. Clearly, such a theorem is proved by providing a simulation, i.e., we want to modify $\mathcal{K}$ in a way such that the stack is emptied whenever $\mathcal{K}$ reaches a final state. In order to do so, we introduce a new state q_λ and a special stack symbol X_0 to avoid acceptance if $\mathcal{K}$ has emptied its stack without having reached a final state. Formally, we proceed as follows.

Let $\mathcal{K} = [Q, \Sigma, \Gamma, \delta, q_0, k_0, F]$ be any given PDA such that $L = L(\mathcal{K})$. We have to construct a PDA $\widetilde{\mathcal{K}}$ such that $L = N(\widetilde{\mathcal{K}})$. We set

$\widetilde{\mathcal{K}} = [Q \cup \{q_\lambda, \tilde{q}_0\}, \Sigma, \Gamma \cup \{X_0\}, \tilde{q}_0, X_0, \tilde{\delta}, \emptyset]$, with $q_\lambda, \tilde{q}_0 \notin Q$, $X_0 \notin \Gamma$, and where $\tilde{\delta}$ is defined as follows:

$\tilde{\delta}(\tilde{q}_0, \lambda, X_0) = \{(q_0, k_0 X_0)\}$,

$\tilde{\delta}(q, a, Z) = \delta(q, a, Z)$ for all $q \in Q \setminus F$, $a \in \Sigma \cup \{\lambda\}$, and $Z \in \Gamma$,

$\tilde{\delta}(q, a, Z) = \delta(q, a, Z)$ for all $q \in F$, $a \in \Sigma$ and $Z \in \Gamma$,

$\tilde{\delta}(q, \lambda, Z) = \delta(q, \lambda, Z) \cup \{(q_\lambda, \lambda)\}$ for all $q \in F$ and $Z \in \Gamma \cup \{X_0\}$,

$\tilde{\delta}(q_\lambda, \lambda, Z) = \{(q_\lambda, \lambda)\}$ for all $Z \in \Gamma \cup \{X_0\}$.

By construction, when starting $\widetilde{\mathcal{K}}$, it is entering the initial instantaneous description of $\mathcal{K}$ but pushes additionally its own stack symbol X_0 into the stack. Then $\widetilde{\mathcal{K}}$ simulates $\mathcal{K}$ until it reaches a final state. If $\mathcal{K}$ reaches a final state, then $\widetilde{\mathcal{K}}$ can either continue to simulate $\mathcal{K}$ or it can change its state to q_λ. If $\widetilde{\mathcal{K}}$ is in q_λ, it can empty the stack and thus accept the input.

Hence, formally we can continue as follows. Let $x \in L(\mathcal{K})$. Then, there is a computation such that

$$(q_0, x, k_0) \xrightarrow[\mathcal{K}]{*} (q, \lambda, \gamma) \quad \text{for a } q \in F .$$

We consider $\widetilde{\mathcal{K}}$ on input x. By its definition, $\widetilde{\mathcal{K}}$ starts in state $\tilde{q}_0$ and with stack symbol X_0. Thus, by using the first spontaneous transition, we get

$$(\tilde{q}_0, x, X_0) \xrightarrow[\widetilde{\mathcal{K}}]{1} (q_0, x, k_0 X_0) .$$

Next, $\widetilde{\mathcal{K}}$ can simulate every step of $\mathcal{K}$'s work; hence we also have

$$(\tilde{q}_0, x, X_0) \xrightarrow[\widetilde{\mathcal{K}}]{1} (q_0, x, k_0 X_0) \xrightarrow[\widetilde{\mathcal{K}}]{*} (q, \lambda, \gamma X_0)$$

Finally, using the last two transitions in the definition of $\tilde{\delta}$ we obtain

$$(q_0, x, k_0 X_0) \xrightarrow[\widetilde{\mathcal{K}}]{*} (q_\lambda, \lambda, \lambda) .$$

Therefore we can conclude that $x \in N(\widetilde{\mathcal{K}})$, and consequently $L(\mathcal{K}) \subseteq N(\widetilde{\mathcal{K}})$. The direction $N(\widetilde{\mathcal{K}}) \subseteq L(\mathcal{K})$ is left as an exercise. ∎

9.1 Introducing Pushdown Automata

So, we have shown that, if a language is accepted by a PDA via final state then it can also be accepted by a PDA via empty stack. It is only natural to ask whether or not the converse is also true. The affirmative answer is given by our next theorem.

Theorem 9.2 *Let* $L = N(\mathcal{K})$ *for a PDA* $\mathcal{K}$. *Then there exists a PDA* $\widetilde{\mathcal{K}}$ *such that* $L = L(\widetilde{\mathcal{K}})$.

Proof. Again the proof is done by simulation. The PDA $\widetilde{\mathcal{K}}$ will simulate $\mathcal{K}$ until it detects that $\mathcal{K}$ has emptied its stack. If this happens then $\widetilde{\mathcal{K}}$ will enter a final state and stop.

The formal construction is as follows. Let $\mathcal{K} = [Q, \Sigma, \Gamma, \delta, q_0, k_0, \emptyset]$ be any given PDA such that $L = N(\mathcal{K})$. We have to construct a PDA $\widetilde{\mathcal{K}}$ such that $L = L(\widetilde{\mathcal{K}})$. We set

$$\widetilde{\mathcal{K}} = [Q \cup \{\tilde{q}_0, q_f\}, \Sigma, \Gamma \cup \{X_0\}, \tilde{\delta}, \tilde{q}_0, X_0, \{q_f\}], \text{ with } \tilde{q}_0, q_f \notin Q, X_0 \notin \Gamma,$$

and where $\tilde{\delta}$ is defined as follows.

$$\tilde{\delta}(\tilde{q}_0, \lambda, X_0) = \{(q_0, k_0 X_0)\},$$
$$\tilde{\delta}(q, a, Z) = \delta(q, a, Z) \quad \text{for all } q \in Q, \ a \in \Sigma \cup \{\lambda\} \text{ and } Z \in \Gamma,$$
$$\tilde{\delta}(q, \lambda, X_0) = \{(q_f, \lambda)\} \quad \text{for all } q \in Q.$$

The first line in the definition of $\tilde{\delta}$ ensures that $\widetilde{\mathcal{K}}$ can start the simulation of $\mathcal{K}$. Note, however, that $\widetilde{\mathcal{K}}$ is putting its own stack symbol *below* the stack symbol of $\mathcal{K}$. The second line in the definition of $\tilde{\delta}$ allows that $\widetilde{\mathcal{K}}$ can simulate all steps of $\mathcal{K}$. If $\mathcal{K}$ empties its stack, then $\widetilde{\mathcal{K}}$ also removes all symbols from its stack except its own stack symbol X_0 while performing the simulation. Finally, the last line in the definition of $\tilde{\delta}$ guarantees that $\widetilde{\mathcal{K}}$ can perform a spontaneous transition into its final state q_f. Thus, $\widetilde{\mathcal{K}}$ then also accepts the input string. The formal verification of $L(\widetilde{\mathcal{K}}) = N(\mathcal{K})$ is left as exercise. ∎

Exercise 9.2 *Complete the proof of Theorem 9.2 by formally showing that* $L(\tilde{\mathcal{K}}) = N(\mathcal{K})$.

9.2 PDAs and Context-Free Languages

So far, we have only dealt with PDAs and their acceptance behavior. It remains to clarify what languages are accepted by PDAs. This is done by the following theorem. Recall that a derivation is said to be a leftmost derivation if at each step in the derivation a production is applied to the leftmost nonterminal.

Theorem 9.3 *Let* $\mathcal{K} = [Q, \Sigma, \Gamma, \delta, q_0, k_0, \emptyset]$ *be any PDA and* $L = N(\mathcal{K})$. *Then* L *is context-free.*

Proof. Let $\mathcal{K} = [Q, \Sigma, \Gamma, \delta, q_0, k_0, \emptyset]$ be any PDA. For proving that L defined as $L = N(\mathcal{K})$ is context-free, we have to construct a context-free grammar $\mathcal{G}$ such that $L = L(\mathcal{G})$. We set $\mathcal{G} = [\Sigma, N, \sigma, P]$, where N is defined as follows. The elements of N are denoted by $[q, A, p]$, where $p, q \in Q$ and $A \in \Gamma$. Additionally, N contains the symbol σ. Next, we have to define the set of productions. P contains the following rules.

(1) $\sigma \to [q_0, k_0, q]$ for every $q \in Q$,

(2) $[q, A, q_{m+1}] \to a[q_1, B_1, q_2][q_2, B_2, q_3] \cdots [q_m, B_m, q_{m+1}]$ for all $q_1, \ldots, q_{m+1} \in Q$ and $A, B_1, \ldots, B_m \in \Gamma$ such that $(q_1, B_1 B_2 \cdots B_m) \in \delta(q, a, A)$ provided $m > 0$.

If $m = 0$ then the production is $[q, A, q_1] \to a$.

To understand the proof it helps to know that the nonterminals and productions of $\mathcal{G}$ have been defined in a way such that a leftmost derivation in $\mathcal{G}$ of a string x is a simulation of the PDA $\mathcal{K}$ when fed the input x. In particular, the nonterminals that appear in any step of a leftmost derivation

9.2 PDAs and Context-Free Languages

in $\mathcal{G}$ correspond to the symbols on the stack of $\mathcal{K}$ at a time when $\mathcal{K}$ has seen as much of the input as the grammar has already generated. In other words, our intention is that $[q, A, p]$ derives x if and only if x causes $\mathcal{K}$ to erase an A from its stack by some sequence of moves beginning in state q and ending in state p.

For showing that $L(\mathcal{G}) = N(\mathcal{K})$ we prove inductively

$$[q, A, p] \underset{\mathcal{G}}{\overset{*}{\Longrightarrow}} x \quad \text{if and only if} \quad (q, x, A) \underset{\mathcal{K}}{\overset{*}{\longrightarrow}} (p, \lambda, \lambda) \,. \tag{9.1}$$

First, we show by induction on i that

$$\text{if} \quad (q, x, A) \underset{\mathcal{K}}{\overset{i}{\longrightarrow}} (p, \lambda, \lambda) \quad \text{then} \quad [q, A, p] \underset{\mathcal{G}}{\overset{*}{\Longrightarrow}} x \,.$$

For the induction basis let $i = 1$. In order to have $(q, x, A) \underset{\mathcal{K}}{\overset{1}{\longrightarrow}} (p, \lambda, \lambda)$ it must hold that $(p, \lambda) \in \delta(q, x, A)$. Consequently, either we have $x = \lambda$ or $x \in \Sigma$. In both cases, by construction of P, we have $([q, A, p] \to x) \in P$. Hence, $[q, A, p] \underset{\mathcal{G}}{\overset{1}{\Longrightarrow}} x$. This proves the induction basis.

Now suppose $i > 0$. Let $x = ay$ and

$$(q, ay, p) \underset{\mathcal{K}}{\overset{1}{\longrightarrow}} (q_1, y, B_1 B_2 \cdots B_n) \underset{\mathcal{K}}{\overset{i-1}{\longrightarrow}} (p, \lambda, \lambda) \,.$$

The string y can be written as $y = y_1 y_2 \cdots y_n$, where y_j has the effect of popping B_j from the stack, possibly after a long sequence of moves. That is, let y_1 be the prefix of y at the end of which the stack first becomes as short as $n - 1$ symbols. Let y_2 be the symbols of y following y_1 such that at the end of y_2 the stack first becomes as short as $n - 2$ symbols, and so on. This arrangement is displayed in Figure 9.2.

Note that B_1 does not need to be the nth stack symbol from the bottom during the entire time y_1 is being read by $\mathcal{K}$, since B_1 may be changed if it is at the top of the stack and is replaced by one or more symbols. However, none of $B_2 B_3 \cdots B_n$ are ever on top while y_1 is being read. Thus,

9. Pushdown Automata

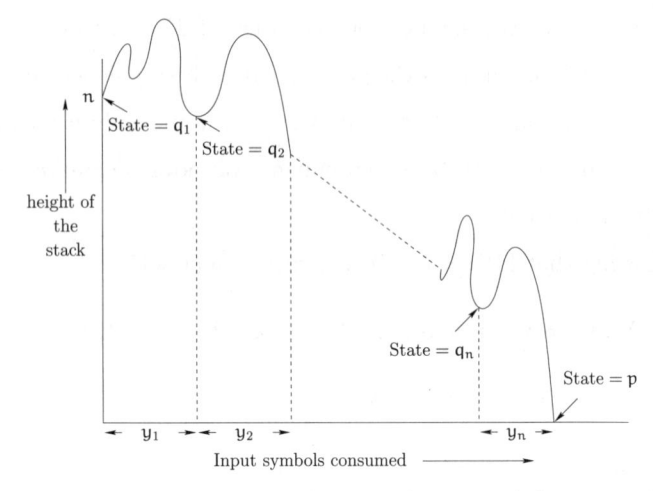

Figure 9.2 Hight of stack as a function of input symbols consumed.

none of $B_2 B_3 \cdots B_n$ can be changed or influence the computation while y_1 is processed. In general, B_j remains on the stack unchanged while $y_1 \cdots y_{j-1}$ is read.

There exists states $q_2, q_3, \ldots, q_{n+1}$, where $q_{n+1} = p$ such that

$$(q_j, y_j, B_j) \xrightarrow[\mathcal{K}]{*} (q_{j+1}, \lambda, \lambda)$$

by fewer than i moves. Note that q_j is the state entered when the stack first becomes as short as $n - j + 1$. Thus, we can apply the induction hypothesis and obtain

$$[q_j, B_j, q_{j+1}] \xRightarrow[g]{*} y_j \quad \text{for } 1 \leq j \leq n.$$

Recalling the original move $(q, ay, p) \xrightarrow[\mathcal{K}]{1} (q_1, y, B_1 B_2 \cdots B_n)$ we know that

$$[q, A, p] \Rightarrow a[q_1, B_1, q_2][q_2, B_2, q_3] \cdots [q_n, B_n, q_{n+1}],$$

thus $[q, A, p] \xRightarrow[g]{*} a y_1 y_2 \cdots y_n = x$, and the sufficiency of (9.1) is shown.

For proving the necessity of (9.1), suppose $[a, A, p] \xRightarrow[g]{i} x$. We prove by induction on i that $(q, x, A) \xrightarrow[\mathcal{K}]{*} (p, \lambda, \lambda)$. The induction basis is for $i = 1$. If $[a, A, p] \xRightarrow[\mathcal{G}]{1} x$, then $([a, A, p] \to x) \in P$ and thus $(p, \lambda) \in \delta(q, x, A)$.

Next, for the induction step suppose

$$[q, A, p] \Rightarrow a[q_1, B_1, q_2] \cdots [q_n, B_n, q_{n+1}] \xRightarrow[\mathcal{G}]{i-1} x ,$$

where $q_{n+1} = p$. We write x as $x = ax_1 \cdots x_n$, where $[q_j, B_j, q_{j+1}] \xRightarrow[\mathcal{G}]{*} x_j$ for $j = 1, \ldots, n$. Moreover, each derivation takes fewer than i steps. Thus, we can apply the induction hypothesis and obtain

$$(q_j, x_j, B_j) \xrightarrow[\mathcal{K}]{*} (q_{j+1}, \lambda, \lambda) \text{ for } j = 1, \ldots, n .$$

If we insert $B_{j+1} \cdots B_n$ at the bottom of each stack in the above sequence of instantaneous descriptions, we see that

$$(q_j, x_j, B_j B_{j+1} \cdots B_n) \xrightarrow[\mathcal{K}]{*} (q_{j+1}, \lambda, B_{j+1} \cdots B_n) . \tag{9.2}$$

Furthermore, from the first step in the derivation of x from $[q, A, p]$ we know that

$$(q, x, A) \xrightarrow[\mathcal{K}]{1} (q_1, x_1 x_2 \cdots x_n, B_1 B_2 \cdots B_n)$$

is a legal move of $\mathcal{K}$. Therefore, from this move and from (9.2) for $j = 1, 2, \ldots, n$ we directly obtain

$$(q, x, A) \xrightarrow[\mathcal{K}]{*} (p, \lambda, \lambda) .$$

This proves the necessity of (9.1).

Finally, we observe that (9.1) with $q = q_0$ and $A = k_0$ says

$$[q_0, k_0, p] \xRightarrow[\mathcal{G}]{*} x \quad \text{if and only if} \quad (q_0, x, k_0) \xrightarrow[\mathcal{K}]{*} (p, \lambda, \lambda) .$$

This observation together with rule (1) of the construction of P says that $\sigma \xRightarrow[\mathcal{G}]{*} x$ if and only if $(q_0, x, k_0) \xrightarrow[\mathcal{K}]{*} (p, \lambda, \lambda)$ for some state p. Therefore, we finally arrive at $x \in L(\mathcal{G})$ if and only if $x \in N(\mathcal{K})$. ∎

Finally, we can state the counterpart of the Chomsky-Schützenberger theorem in terms of PDAs. For proving it, please look at any accepting computation of a PDA. Consider the sequence of read and write operations on the stack. If a symbol Z is read from the stack, write down $\overline{a}_Z$ and if a symbol Z is written, then write down a_Z. This should result in an element of a Dyck language.

Exercise 9.3 *For every PDA $\mathcal{K}$ there exists a number n, homomorphisms g, h and a regular language $R_\mathcal{K}$ such that* $L(\mathcal{K}) = h(g^{-1}(D_n) \cap R_\mathcal{K})$.

9.3 Problem Set 9

Problem 9.1 Show that for deterministic PDAs acceptance via empty stack is *weaker* than acceptance via final state. Discuss the insight obtained.

Problem 9.2 Prove or disprove the following theorem.

Theorem 9.4 *For all* $L \in \mathcal{CF}$ *and for all* $R \in \mathcal{REG}$ *we have* $L \cap R \in \mathcal{CF}$.

Problem 9.3 Consider the following language

$$L = \{0^n 1^m \mid n, m \in \mathbb{N},\ n \leq m \leq 2n\}\ .$$

(1) Prove that L is context-free.

(2) Construct a PDA $\mathcal{K} = [Q, \Sigma, \Gamma, \delta, q_0, k_0, F]$ such that $L(\mathcal{K}) = L$.

10
$\mathcal{CF}$, PDAs and Beyond

We want to show that all context-free languages are accepted by pushdown automata. For doing this, it is very convenient to use another normal form for context-free grammars, i.e., the so-called Greibach [11] normal form (abbr. GNF) (グライバッハ標準形).

10.1 Greibach Normal Form

Definition 10.1 (Greibach Normal Form) *A context-free grammar* $\mathcal{G} = [\mathsf{T}, \mathsf{N}, \sigma, \mathsf{P}]$ *is said to be in* Greibach normal form *if all productions of* P *have the form* $\mathsf{h} \to \mathsf{a}\alpha$, *where* $\mathsf{a} \in \mathsf{T}$ *and* $\alpha \in (\mathsf{N} \setminus \{\sigma\})^*$.

We aim to show that every language $\mathsf{L} \in \mathcal{CF}$ possesses a grammar in GNF. This is, however, difficult. Following [16], we need two lemmata and the following notion. For a context-free grammar $\mathcal{G}$, an h-*production* is a production with nonterminal h on the left.

Lemma 10.1 *Let* $\mathcal{G} = [\mathsf{T}, \mathsf{N}, \sigma, \mathsf{P}]$ *be any context-free grammar. Furthermore, let* $\mathsf{h} \to \alpha_1 \mathsf{h}' \alpha_2$ *be any production from* P, *where* $\mathsf{h}, \mathsf{h}' \in \mathsf{N}$ *and let* $\mathsf{h}' \to \beta_1$, $\mathsf{h}' \to \beta_2$, ..., $\mathsf{h}' \to \beta_r$ *be all* h'-*productions. Let* $\mathcal{G}_1 = [\mathsf{T}, \mathsf{N}, \sigma, \mathsf{P}_1]$ *be the grammar obtained from* $\mathcal{G}$ *by deleting the production* $\mathsf{h} \to \alpha_1 \mathsf{h}' \alpha_2$ *and by adding the productions* $\mathsf{h} \to \alpha_1 \beta_1 \alpha_2$, $\mathsf{h} \to \alpha_1 \beta_2 \alpha_2$, ..., $\mathsf{h} \to \alpha_1 \beta_r \alpha_2$. *Then* $\mathsf{L}(\mathcal{G}) = \mathsf{L}(\mathcal{G}_1)$.

Proof. The inclusion $L(\mathcal{G}_1) \subseteq L(\mathcal{G})$ is obvious, since if $h \to \alpha_1 \beta_i \alpha_2$ is used in a derivation of $\mathcal{G}_1$, then

$$h \underset{\mathcal{G}}{\overset{1}{\Longrightarrow}} \alpha_1 h' \alpha_2 \underset{\mathcal{G}}{\overset{1}{\Longrightarrow}} \alpha_1 \beta_i \alpha_2$$

can be used in $\mathcal{G}$.

For the opposite direction $L(\mathcal{G}) \subseteq L(\mathcal{G}_1)$, one notes that $h \to \alpha_1 h' \alpha_2$ is the only production which is in $\mathcal{G}$ but not in $\mathcal{G}_1$. Whenever this production is used in a derivation by $\mathcal{G}$, the nonterminal h' must be rewritten at some late step by using a production of the form $h' \to \beta_i$ for some $i \in \{1, \ldots, r\}$. These two steps can be replaced by the single step $h' \underset{\mathcal{G}_1}{\overset{1}{\Longrightarrow}} \alpha_1 \beta_i \alpha_2$. ∎

Lemma 10.2 *Let $\mathcal{G} = [T, N, \sigma, P]$ be any context-free grammar. Let $h \to h\alpha_1, \ldots, h \to h\alpha_r$ be all h-productions for which h is the leftmost symbol of the right-hand side. Furthermore, let $h \to \beta_1, \ldots, h \to \beta_s$ be the remaining h-productions. Let $\mathcal{G}_1 = [T, N \cup \{B\}, \sigma, P_1]$ be the context-free grammar formed by adding the nonterminal B to N and by replacing all the h-productions by $h \to \beta_i$ and $h \to \beta_i B$ for $i = 1, \ldots, s$ and all the remaining h-productions by $B \to \alpha_j$ and $B \to \alpha_j B$, $j = 1, \ldots, r$. Then $L(\mathcal{G}) = L(\mathcal{G}_1)$.*

We leave it as an exercise to show Lemma 10.1.

We continue by providing the following fundamental theorem.

Theorem 10.1 *For every language $L \in \mathcal{CF}$ with $\lambda \notin L$ there exists a grammar $\tilde{\mathcal{G}}$ such that $L = L(\tilde{\mathcal{G}})$ and $\tilde{\mathcal{G}}$ is in Greibach normal form.*

Proof. Let L be any context-free language with $\lambda \notin L$. Then there exists a grammar $\mathcal{G} = [T, N, \sigma, P]$ in CNF such that $L = L(\mathcal{G})$ (cf. Theorem 7.2). Furthermore, let $N = \{h_1, h_2, \ldots, h_m\}$. The first step in the construction of the GNF is to modify the productions of $\mathcal{G}$ in a way such that if $h_i \to h_j \gamma$ is a production, then $j > i$. Starting with h_1 and proceeding to h_m this is done as follows.

10.1 Greibach Normal Form

Assume that the productions have been modified so that for $1 \leq i < k$, $h_i \to h_j \gamma$ is a production only if $j > i$. Now, we are modifying the h_k-productions.

If $h_k \to h_j \gamma$ is a production with $j < k$, then we generate a new set of productions by substituting for h_j the right-hand side of each h_j-production according to Lemma 10.1. By repeating the process at most $k - 1$ times, we obtain productions of the form $h_k \to h_\ell \gamma$, where $\ell \geq k$. The productions with $\ell = k$ are then replaced by applying Lemma 10.2. That means we have to introduce a new nonterminal B_k. The complete algorithm is provided in Figure 10.1.

```
     begin
(1)    for k := 1 to m do
          begin
(2)          for j := 1 to k − 1 do
(3)             for each production of the form h_k → h_j α do
                   begin
(4)                   for all productions h_j → β do
(5)                      add production h_k → β α;
(6)                   remove production h_k → h_j α
                   end;
(7)             for each production of the form h_k → h_k α do
                   begin
(8)                   add productions B_k → α and B_k → α B_k;
(9)                   remove production h_k → h_k α
                   end;
(10)            for each production h_k → β, where β does not
                    begin with h_k do
(11)               add production h_k → β B_k
          end
     end
```

Figure 10.1 Step 1 in the Greibach normal form algorithm.

By repeating the above process for each original variable, we have only productions of the forms:

1) $h_i \to h_j \gamma$, $j > i$,
2) $h_j a \gamma$, $a \in T$,
3) $B_i \to \gamma$, $\gamma \in (N \cup \{B_1, B_2, \ldots, B_{j-1}\})^*$.

Note that the leftmost symbol on the right-hand side of any production for h_m must be a terminal, since h_m is the highest-numbered nonterminal. The leftmost symbol on the right-hand side of any production for h_{m-1} must be either h_m or a terminal symbol. When it is h_m, we can generate new productions by replacing h_m by the right-hand side of the productions for h_m according to Lemma 10.1. These productions must have right-hand sides that start with a terminal symbol. We then proceed to the productions for $h_{m-2}, \ldots, h_2, h_1$ until the right hand-side of each production for an h_i starts with a terminal symbol.

As the last step we examine the productions of the new nonterminals $B_1, \ldots, B_m$. Since we started with a grammar in CNF, it is easy to prove by induction on the number of applications of Lemmata 10.1 and 10.2 that the right-hand side of every h_i-production, $1 \le i \le n$, begins with a terminal or $h_j h_k$ for some j and k. Thus α in Instruction (7) of Figure 10.1 can never be empty or begin with some B_j. So no B_i-production can start with another B_j. Therefore, all B_i-productions have right hand sides beginning with terminals or h_i's, and one more application of Lemma 10.1 for each B_i-production completes the construction. ∎

Since the construction outlined above is rather complicated, we exemplify it by looking at the following grammar $\mathcal{G}$.

Example 10.1 *Let* $\mathcal{G} = [\{a, b\}, \{h_1, h_2, h_3\}, h_1, P]$, *where*

$$P = \{h_1 \to h_2 h_3,\ h_2 \to h_3 h_1,\ h_2 \to b,\ h_3 \to h_1 h_2,\ h_3 \to a\}.$$

We want to convert $\mathcal{G}$ into GNF.

10.1 Greibach Normal Form

Step 1. Since the right-hand side of the productions for h_1 and h_2 start with terminals or higher numbered nonterminals, we begin with the production $h_3 \to h_1 h_2$ and substitute the string $h_2 h_3$ for h_1. Note that $h_1 \to h_2 h_3$ is the only production with h_1 on the left.

The resulting set of productions is:

$h_1 \to h_2 h_3$ $\qquad$ $h_2 \to b$ $\qquad$ $h_3 \to a$

$h_2 \to h_3 h_1$ $\qquad$ $h_3 \to h_2 h_3 h_2$

Since the right-hand side of the production $h_3 \to h_2 h_3 h_2$ begins with a lower numbered nonterminal, we substitute for the first occurrence of h_2 both $h_3 h_1$ and b. Thus, $h_3 \to h_2 h_3 h_2$ is replaced by $h_3 \to h_3 h_1 h_3 h_2$ and $h_3 \to b h_3 h_2$. The new set of productions is:

$h_1 \to h_2 h_3$ $\qquad$ $h_2 \to b$ $\qquad$ $h_3 \to b h_3 h_2$

$h_2 \to h_3 h_1$ $\qquad$ $h_3 \to h_3 h_1 h_3 h_2$ $\qquad$ $h_3 \to a$

At this point, we apply Lemma 10.2 to the productions $h_3 \to h_3 h_1 h_3 h_2$, $h_3 \to b h_3 h_2$ and $h_3 \to a$.

Symbol B_3 is introduced, and the production $h_3 \to h_3 h_1 h_3 h_2$ is replaced by $h_3 \to b h_3 h_2 B_3$, $h_3 \to a B_3$, $B_3 \to h_1 h_3 h_2 B_3$, and $B_3 \to h_1 h_3 h_2 B_3$. For simplifying notation, we adopt the BNF. The resulting set of productions is:

$h_1 \to h_2 h_3$ $\qquad\qquad$ $h_3 \to b h_3 h_2 B_3 \mid a B_3 \mid b h_3 h_2 \mid a$

$h_2 \to h_3 h_1 \mid b$ $\qquad\qquad$ $B_3 \to h_1 h_3 h_2 B_3 \mid h_1 h_3 h_2 B_3$

Step 2. Now all the productions with h_3 on the left have right-hand sides that start with terminals. These are used to replace h_3 in the production

$h_2 \to h_3 h_1$ and then the productions with h_2 on the left are used to replace h_2 in the production $h_1 \to h_2 h_3$. The result is the following:

$h_3 \to bh_3h_2B_3 \mid bh_3h_2 \mid aB_3 \mid a$

$h_2 \to bh_3h_2B_3h_1 \mid bh_3h_2h_1 \mid aB_3h_1 \mid ah_1 \mid b$

$h_1 \to bh_3h_2B_3h_1h_3 \mid bh_3h_2h_1h_3 \mid aB_3h_1h_3 \mid ah_1h_3 \mid bh_3$

$B_3 \to h_1h_3h_2 \mid h_1h_3h_2B_3$.

Step 3. The two B_3-productions are converted to proper form, resulting in 10 more productions. That is, the productions $B_3 \to h_1h_3h_2$ and $B_3 \to h_1h_3h_2B_3$ are altered by substituting the right side of each of the five productions with h_1 on the left for the first occurrences of h_1. Thus, $B_3 \to h_1h_3h_2$ becomes

$B_3 \to bh_3h_2B_3h_1h_3h_3h_2 \mid aB_3h_1h_3h_3h_2 \mid bh_3h_3h_2$

$\mid bh_3h_2h_1h_3h_3h_2 \mid ah_1h_3h_3h_2$.

The other production for B_3 is replaced similarly. The final set of productions is thus:

$h_3 \to bh_3h_2B_3 \mid bh_3h_2 \mid aB_3 \mid a$

$h_2 \to bh_3h_2B_3h_1 \mid bh_3h_2h_1 \mid aB_3h_1 \mid ah_1 \mid b$

$h_1 \to bh_3h_2B_3h_1h_3 \mid bh_3h_2h_1h_3 \mid aB_3h_1h_3 \mid ah_1h_3 \mid bh_3$

$B_3 \to bh_3h_2B_3h_1h_3h_3h_2 \mid aB_3h_1h_3h_3h_2 \mid bh_3h_3h_2$

$\mid bh_3h_2h_1h_3h_3h_2 \mid ah_1h_3h_3h_2$

$\mid bh_3h_2B_3h_1h_3h_3h_2B_3 \mid aB_3h_1h_3h_3h_2B_3 \mid bh_3h_3h_2B_3$

$\mid bh_3h_2h_1h_3h_3h_2B_3 \mid ah_1h_3h_3h_2B_3$.

end (Example)

10.2 Main Theorem

Now, we are in the position to show the remaining fundamental theorem concerning the power of pushdown automata.

Theorem 10.2 *For every language* $L \in \mathcal{CF}$ *there exists a PDA* $\mathcal{K}$ *such that* $L = N(\mathcal{K})$.

Proof. We assume that $\lambda \notin L$. It is left as an exercise to modify the construction for the case that $\lambda \in L$. Let $\mathcal{G} = [T, N, \sigma, P]$ be a context-free grammar in GNF such that $L = L(\mathcal{G})$. Furthermore, let

$$\mathcal{K} = [\{q\}, T, N, \delta, q, \sigma, \emptyset] \,, \tag{10.1}$$

where $(q, \gamma) \in \delta(q, a, A)$ whenever $(A \to a\gamma) \in P$.

The PDA $\mathcal{K}$ simulates leftmost derivations of $\mathcal{G}$. Since $\mathcal{G}$ is in GNF, each sentential form[†] in a leftmost derivation consists of a string x of terminals followed by a string of nonterminals α. The PDA $\mathcal{K}$ stores the suffix α of the left sentential form on its stack after processing the prefix x. Formally, we show the following claim.

Claim 1. $\sigma \underset{\mathcal{G}}{\overset{*}{\Longrightarrow}} x\alpha$ *by a leftmost derivation iff* $(q, x, \sigma) \underset{\mathcal{K}}{\overset{*}{\longrightarrow}} (q, \lambda, \alpha)$.

We start with the sufficiency. The prove is done by induction. That is, we assume $(q, x, \sigma) \underset{\mathcal{K}}{\overset{i}{\longrightarrow}} (q, \lambda, \alpha)$ and show $\sigma \underset{\mathcal{G}}{\overset{*}{\Longrightarrow}} x\alpha$.

The induction basis is for $i = 0$. Thus, we assume $(q, x, \sigma) \underset{\mathcal{K}}{\overset{0}{\longrightarrow}} (q, \lambda, \alpha)$. By the definition of the reflexive-transitive closure $\underset{\mathcal{K}}{\overset{*}{\longrightarrow}}$, this means nothing else than $(q, x, \sigma) = (q, \lambda, \alpha)$. Consequently, $x = \lambda$ and $\alpha = \sigma$. Obviously, by the definition of the reflexive-transitive closure $\underset{\mathcal{G}}{\overset{*}{\Longrightarrow}}$, we can conclude $\sigma \underset{\mathcal{G}}{\overset{*}{\Longrightarrow}} \sigma$, again in zero steps. This proves the induction basis.

[†] A string α of terminals and nonterminals is called a *sentential form* (文形式) if $\sigma \overset{*}{\Rightarrow} \alpha$.

For the induction step, assume $i \geq 1$ and let $x = ya$, where $y \in T^*$. Now, we consider the next-to-last-step, i.e.,

$$(q, ya, \sigma) \xrightarrow[\mathcal{K}]{i-1} (q, a, \beta) \xrightarrow[\mathcal{K}]{1} (q, \lambda, \alpha) . \tag{10.2}$$

If we remove a from the end of the input string in the first i instantaneous descriptions of the sequence (10.2), we see that $(q, y, \sigma) \xrightarrow[\mathcal{K}]{i-1} (q, \lambda, \beta)$, since a cannot influence $\mathcal{K}$'s behavior until it is actually consumed from the input. Thus, we can apply the induction hypothesis and obtain

$$\sigma \xRightarrow[\mathcal{G}]{*} y\beta . \tag{10.3}$$

Taking into account that the pushdown automaton $\mathcal{K}$, while consuming a, is making the move $(q, a, \beta) \xrightarrow[\mathcal{K}]{1} (q, \lambda, \alpha)$, we directly get by construction that $\beta = A\gamma$ for some $A \in N$, $(A \to a\eta) \in P$ and $\alpha = \eta\gamma$. Hence, combining the latter with (10.3), we arrive at

$$\sigma \xRightarrow[\mathcal{G}]{*} y\beta \xRightarrow[\mathcal{G}]{1} ya\eta\gamma = x\alpha .$$

This completes the sufficiency proof.

For showing the necessity, suppose that $\sigma \xRightarrow[\mathcal{G}_*]{i} x\alpha$ by a leftmost derivation. We prove by induction on i that $(q, x, \sigma) \xrightarrow[\mathcal{K}]{} (q, \lambda, \alpha)$. The induction basis is again done for $i = 0$, and can be shown by using similar arguments as above.

For the induction step, let $i \geq 1$ and suppose

$$\sigma \xRightarrow[\mathcal{G}]{i-1} yA\gamma \xRightarrow[\mathcal{G}]{1} ya\eta\gamma ,$$

where $x = ya$ and $\alpha = \eta\gamma$. By the induction hypothesis, we directly get

$$(q, y, \sigma) \xrightarrow[\mathcal{K}]{*} (q, \lambda, A\gamma)$$

and thus $(q, ya, \sigma) \xrightarrow[\mathcal{K}]{*} (q, a, A\gamma)$. Since $(A \to a\eta) \in P$, we can conclude that $(q, \eta) \in \delta(q, a, A)$. Therefore,

$$(q,x,\sigma) \xrightarrow[\mathcal{K}]{*} (q,a,A\gamma) \xrightarrow[\mathcal{K}]{1} (q,\lambda,\alpha)$$

and the necessity follows. This proves Claim 1.

To conclude the proof of the theorem, we have only to note that Claim 1 with $\alpha = \lambda$ says

$$\sigma \xRightarrow[\mathcal{G}]{*} x \quad \text{if and only if} \quad (q,x,\sigma) \xrightarrow[\mathcal{K}]{*} (q,\lambda,\lambda) \ .$$

That is, $x \in L(\mathcal{G})$ if and only if $x \in N(\mathcal{K})$. ∎

Theorems 9.3 and 10.2 as well as Theorems 9.1 and 9.2 together directly allow for the following main theorem.

Theorem 10.3 *Let L be any language. Then the following three assertions are equivalent:*

(1) $L \in \mathcal{CF}$.

(2) *There exists a pushdown automaton $\mathcal{K}_1$ such that $L = L(\mathcal{K}_1)$.*

(3) *There exists a pushdown automaton $\mathcal{K}_2$ such that $L = N(\mathcal{K}_2)$.*

After so much progress we may want to ask questions like whether or not $L(\mathcal{G}_1) \cap L(\mathcal{G}_2) = \emptyset$ for any given context-free grammars $\mathcal{G}_1$, $\mathcal{G}_2$. Remembering Corollary 6.2 we may be tempted to think this is not a too difficult task. But as a matter of fact, nobody succeeded to design an algorithm solving this problem for all context-free grammars. Maybe, there is a deeper reason behind this situation. Before we can explore such problems, we have to deal with *computability* (計算可能性).

On the other hand, so far we have studied regular and context-free languages. But we have already seen a language which is not context-free, i.e., $L = \{a^n b^n c^n \mid n \in \mathbb{N}\}$ (cf. Theorem 7.4). Thus, it is only natural to ask what other language families are around. Due to the lack of space, we can only sketch these parts of formal language theory.

10.3 Context-Sensitive Languages

First, we provide a formal definition.

Definition 10.2 *A Grammar* $\mathcal{G} = [T, N, \sigma, P]$ *is said to be* context-sensitive (文脈依存文法) *if all productions satisfy the following conditions.*

(1) $(\alpha \to \beta) \in P$ *iff there are* $s_1, s_2, r,$ *and* h *such that* $h \in N$, $s_1, s_2 \in (T \cup N)^*$ *and* $r \in (T \cup N)^+$ *and* $\alpha = s_1 h s_2$ *and* $\beta = s_1 r s_2$, *or*

(2) $\alpha = h$ *and* $\beta = \lambda$ *and* h *does not occur at any right-hand side of a production from* P.

Definition 10.3 *A language is said to be* context-sensitive (文脈依存言語) *if there exists a context-sensitive grammar* $\mathcal{G}$ *such that* $L = L(\mathcal{G})$.

By $\mathcal{CS}$ we denote the family of all context-sensitive languages. The name context-sensitive is quite intuitive, since the replacement or rewriting of a nonterminal is only possible in a certain context expressed by a prefix s_1 and suffix s_2. Definition 10.2 directly allows for the observation that $\mathcal{CF} \subseteq \mathcal{CS}$.

Exercise 10.1 *Prove that* $\mathcal{CF} \subset \mathcal{CS}$.

Furthermore, using the same ideas *mutatis mutandis* as in the proof of Theorem 6.2 one can easily show the following.

Theorem 10.4 *The context-sensitive languages are closed under union, product and Kleene closure.*

Also, in the same way as Theorem 6.3 has been shown, one can prove the context-sensitive languages to be closed under transposition. That is, we directly get the next theorem.

Theorem 10.5 *Let* Σ *be any alphabet, and let* $L \subseteq \Sigma^*$. *Then we have: If* $L \in \mathcal{CS}$ *then* $L^T \in \mathcal{CS}$, *too.*

Moreover, in contrast to Theorem 6.4 we have the following.

10.3 Context-Sensitive Languages

Theorem 10.6 $\mathcal{CS}$ *is closed under intersection.*

For establishing further properties of context-sensitive languages, we need the following definition.

Definition 10.4 *A grammar* $\mathcal{G} = [T, N, \sigma, P]$ *is said to be* length-increasing (非縮小文法) *if each production* $(\alpha \to \beta) \in P$ *satisfies the condition* $|\alpha| \le |\beta|$. *In addition,* $(\sigma \to \lambda) \in P$ *is allowed, provided that in this case* σ *does not occur on the right-hand side of any production from* P.

Looking at Definition 10.2 directly allows for the following corollary.

Corollary 10.1 *Every context-sensitive grammar is length-increasing.*

The opposite is not true (cf. Example 10.2). Nevertheless, one can show the following.

Theorem 10.7 *For every length-increasing grammar there exists an equivalent context-sensitive grammar.*

The proof of the latter theorem is also left as an exercise. Putting Corollary 10.1 and Theorem 10.7 together directly yields the following equivalence.

Theorem 10.8 *Let* L *be any language. Then the following statements are equivalent:*

(1) *There exists a context-sensitive grammar* $\mathcal{G}$ *such that* $L = L(\mathcal{G})$.

(2) *There exists a length-increasing grammar* $\tilde{\mathcal{G}}$ *such that* $L = L(\tilde{\mathcal{G}})$.

Example 10.2 Let $T \ne \emptyset$ be any alphabet. We define a grammar $\mathcal{G} = [T, N, \sigma, P]$ as follows. Let

$$N = \{\sigma\} \cup \{X_i \mid i \in T\} \cup \{A_i \mid i \in T\} \cup \{B_i \mid i \in T\},$$

and let P be the following set of productions, where $i, j \in T$:

1. $\sigma \to i\sigma X_i$ 3. $B_i X_j \to X_j B_i$ 5. $A_i X_j \to A_i B_j$
2. $\sigma \to A_i B_i$ 4. $B_i \to i$ 6. $A_i \to i$

Inspecting the productions we see that $\mathcal{G}$ is a length-increasing grammar which is not context-sensitive, since the context in Production 3 is destroyed. Nevertheless, by Theorem 10.7 we know that the language $L(\mathcal{G})$ is context-sensitive. But what language is generated by $\mathcal{G}$? The answer is provided by our next exercise.

Exercise 10.2 *Prove that the grammar $\mathcal{G}$ given in Example 10.2 generates* $L = \{ww \mid w \in T^+\}$.

Exercise 10.3 *Provide a context-sensitive grammar for the language* $L = \{ww \mid w \in T^+\}$.

The notion of length-increasing grammar has another nice implication which we state next.

Theorem 10.9 *There exists an algorithm that on input any context-sensitive grammar $\mathcal{G} = [T, N, \sigma, P]$ and any string $s \in T^*$ decides whether or not $s \in L(\mathcal{G})$.*

Proof. Since every context-sensitive grammar is also a length-increasing grammar, it suffices to examine all finite sequences $w_0, w_1, \ldots, w_n$ with $|w_i| \leq |w_{i+1}|$, $w_i \neq w_{i+1}$, $i = 0, \ldots, n-1$, and $\sigma = w_0$ as well as $w_n = s$, where $w_i \in (T \cup N)^+$. The number of all those sequences is finite. Let $\mathcal{S}$ be the set of all such sequences.

Now, the only thing one has to check is whether or not

$$w_i \underset{\mathcal{G}}{\overset{1}{\Longrightarrow}} w_{i+1} \quad \text{for all } i = 0, \ldots, n-1 \,. \tag{10.4}$$

So, one either finds a sequence in $\mathcal{S}$ fulfilling (10.4). Then one can conclude that $s \in L(\mathcal{G})$. If all sequences in $\mathcal{S}$ fail to satisfy (10.4) then $s \notin L(\mathcal{G})$. ∎

Recall that we have proved the regular languages to be closed under complement (cf. Section 16.4) and the context-free languages to be *not* closed under complement (cf. Theorem 6.4). As curious as we are, we like to know whether or not the context-sensitive languages are closed under complement. To answer this question is by no means easy. As a matter of fact, it took more than 20 years to resolve this problem. The *affirmative* answer is based on a major breakthrough in complexity theory obtained by Immerman [17] and Szelepcsényi [36]. Combining it with Kuroda's [21] characterization of $\mathcal{CS}$ in terms of linearly space bounded nondeterministic Turing machines shows the closure of $\mathcal{CS}$ under complement.

Finally, we mention what happens if we pose no restrictions whatsoever on the set of productions. The resulting family of languages is $\mathcal{L}_0$ (see Chapter 1). Then, it is quite obvious that all languages L in $\mathcal{L}_0$ share the property that we can algorithmically *enumerate all and only* the elements contained in L provided we are given a grammar $\mathcal{G}$ for L. However, as we shall see later, Theorem 10.9 *cannot* be generalized to $\mathcal{L}_0$ (cf. Theorem 14.6). So, we can directly conclude that $\mathcal{CS} \subset \mathcal{L}_0$. Putting it all together gives us the Chomsky hierarchy (チョムスキー階層), i.e.,

$$\mathcal{REG} \subset \mathcal{CF} \subset \mathcal{CS} \subset \mathcal{L}_0.$$

10.4 Problem Set 10

Problem 10.1 Prove Theorem 10.5.

Problem 10.2 Prove or disprove the following theorem.

Theorem 10.10 $\mathcal{CF}$ *is closed under inverse homomorphisms.*

Problem 10.3 Let $\mathcal{G} = [\{a, b\}, \{\sigma, B, K, S, W\}, \sigma, P]$ be given, where

$$P = \{\sigma \to SaK,\ aK \to WbbK,\ aW \to Wbb,\ SWb \to SaB,$$
$$SWb \to aB,\ Bb \to aB,\ BK \to K,\ BK \to \lambda\}.$$

(1) Determine $L(\mathcal{G})$.

(2) Prove the correctness of the assertion you made in (1).

(3) Prove or disprove the grammar $\mathcal{G}$ to be context-sensitive.

11 Models of Computation

The history of algorithms goes back, approximately, to the origins of mathematics at all. For thousands of years, in most cases, the solution of a mathematical problem was equivalent to the design of an algorithm that *solved* it. The ancient development of algorithms culminated in Euclid's *Elements* (ユークリッド原論). For example, Book VII of the elements contains the Euclidean algorithm for computing the greatest common divisor of two integers. The Elements have been studied for 20 centuries in many languages starting, of course, in the original Greek, then in Arabic, Latin, and many modern languages.

A larger part of Euclid's *Elements* deals with the problem to construct geometrical figures by using only ruler and compass. Over the centuries, often quite different constructions have been proposed for certain problems. Moreover, in classical geometry there are also a couple of construction problems that nobody could solve by using only ruler and compass. Perhaps the most famous of these problems are the trisection of an angle, squaring the circle and duplicating the cube. Another important example is the question which regular n-gons are constructible by using only ruler and compass. The latter problem was only resolved by Gauss in 1798.

However, even after Lindemann's proof in 1882 that it is impossible to square the circle, it took roughly another 50 years before modern computability theory started. The main step to be undertaken was to *formal-*

ize the notion of algorithm. The famous impossibility results obtained for the classical geometrical problems "only" proved that there is no particular type of algorithm solving, e.g. the problem to square the circle. Here the elementary operations are the application of ruler and compass.

What else can be said concerning the notion of algorithm? The term *algorithm* is derived from the name of Al-Hwarizmi (approx.: 780 - 850) who worked in the *house of wisdom* in Bagdad. He combined the scientific strength of Greek mathematics with the versatility of Indian mathematics to perform calculations. Another influential source for the development of our thinking was Raimundus Lullus (1232 - 1316) who published more than 280 papers. His *Ars magna* (Engl.: the great art) developed the idea to logically combine termini by using a machine.

Inspired by Lullus' ideas Leibniz (1646 - 1716) split the *Ars magna* into an *Art indicanti* (decision procedures) and an *Art inveniendi* (generation or enumeration procedures). Here by *decision procedure* (決定手続き) a method is meant which, for every object, can *find out* within a *finite* amount of time whether or not it possesses the property asked for.

An *enumeration procedure* (列挙手続き) outputs all and only the objects having the property asked for. Therefore, in a finite amount of time we can only detect that an object has the property asked for, if it has this property. For objects *not* possessing the property asked for, in general, within a finite amount of time we *cannot* detect the absence of the property.

Also, Leibniz pointed out that both decision procedures and enumeration procedures must be realizable by a machine. He also designed a machine for the four basic arithmetic operations and presented it in 1673 at the Royal Society in London. Note that also Schickardt (1624) and Pascal (1641) have built such machines. As a matter of fact, Leibniz was convinced that one can find for any problem an algorithm solving it.

11. Models of Computation

But there have been problems around that could not be solved despite enormous efforts of numerous mathematicians. For example, the design of an algorithm deciding whether a given Diophantine equation has an integral solution (Hilbert's 10th problem) remained unsolved until 1967 when it was shown by Matiyasevich [24] that there is no such algorithm. So, modern computation theory starts with the question:

Which problems can be solved algorithmically ?

In order to answer it, first of all, the *intuitive notion of an algorithm has to be formalized mathematically.*

We assume an intuitive understanding of what an algorithm is. But what is meant by "there is no algorithm solving problem Π," where Π is e.g. Hilbert's 10th problem or the problem stated at the end of Chapter 10? Having a particular algorithm on hand, we can check if it is solving problem Π. However, what can we say about all the algorithms still to be discovered? How can we know that none of them will ever solve problem Π?

We may think of algorithms as of computer programs. There are many computer programs around and many projects under development. While we are sleeping, a new computer program may be written. In general, we have no idea who is writing what computer program. There are very talented programmers. So, how can we know that during the next year, or during the next decades there will be no program solving our problem Π?

This is the point where the beauty and strength of mathematics comes into play. Let us see how we can get started. Looking at all the algorithms we know, we can say that an algorithm is a computation method having the following properties.

(1) The instruction is a finite text.

(2) The computation is done step by step, where each step performs an elementary operation.

(3) In each step of the execution of the computation it is uniquely determined which elementary operation we have to perform.

(4) The next computation step depends only on the input and the intermediate results computed so far.

Now, we can also assume that there is a finite alphabet Σ such that every algorithm can be represented as a string from Σ^*. Since the number of all strings from Σ^* is *countably infinite* there are at most countably infinite many algorithms. By Theorem 1.1 we have that $\{f \mid f\colon \mathbb{N} \to \{0,1\}\}$ is *uncountably infinite* (非可算無限). So, we get the following theorem.

Theorem 11.1 *There exists a noncomputable function* $f\colon \mathbb{N} \to \{0,1\}$.

While this result is of fundamental epistemological importance, it is telling nothing about any particular function. For achieving results in this regard, we have to do much more. We start by formalizing the notion of algorithm in a mathematically precise way (our description given above is not a formal one). There are several ways this formalization can be done and we shall see two of them. First, we present Gödel's [10] approach.

11.1 Partial Recursive Functions

For all $n \in \mathbb{N}^+$ we write $\mathcal{P}^n$ to denote the set of all partial recursive functions (部分帰納的関数) from $\mathbb{N}^n$ into $\mathbb{N}$. Here we define $\mathbb{N}^1 = \mathbb{N}$ and $\mathbb{N}^{n+1} = \mathbb{N}^n \times \mathbb{N}$, i.e., $\mathbb{N}^n$ is the set of all ordered n-tuples of natural numbers. Gödel's [10] approach to define the set $\mathcal{P}$ of all partial recursive functions is as follows.

Step (1): Define some basic functions which are intuitively computable.

Step (2): Define some rules that can be used to construct new computable functions from functions that are already known to be computable.

11. Models of Computation

For Step (1) we define the following functions Z, S, V: $\mathbb{N} \to \mathbb{N}$ by setting

$$Z(n) = 0 \quad \text{for all } n \in \mathbb{N},$$

$$S(n) = n + 1 \quad \text{for all } n \in \mathbb{N},$$

$$V(n) = \begin{cases} 0, & \text{if } n = 0; \\ n - 1, & \text{for all } n \geq 1. \end{cases}$$

That is, Z is the constant 0 function (定数 0 関数), S is the successor function (後者関数) and V is the predecessor function (前者関数). Clearly, these functions are intuitively computable. So, by definition, we have Z, S, V $\in \mathcal{P}^1$. This completes Step (1) of the outline given above. Next, we define the rules (規則)(cf. Step (2)).

(2.1) (Introduction of fictitious variables （架空変数の導入）)

Let $n \in \mathbb{N}^+$; then we have:

if $\tau \in \mathcal{P}^n$ and $\psi(x_1, \ldots, x_n, x_{n+1}) =_{df} \tau(x_1, \ldots, x_n)$, then $\psi \in \mathcal{P}^{n+1}$.

(2.2) (Identifying variables （変数の同一視）)

Let $n \in \mathbb{N}^+$; then we have:

if $\tau \in \mathcal{P}^{n+1}$ and $\psi(x_1, \ldots, x_n) =_{df} \tau(x_1, \ldots, x_n, x_n)$, then $\psi \in \mathcal{P}^n$.

(2.3) (Permuting variables （変数の置換）)

Let $n \in \mathbb{N}^+$, $n \geq 2$, and let $i \in \{1, \ldots, n\}$; then we have: if $\tau \in \mathcal{P}^n$ and $\psi(x_1, \ldots, x_i, x_{i+1}, \ldots, x_n) =_{df} \tau(x_1, \ldots, x_{i+1}, x_i, \ldots, x_n)$, then $\psi \in \mathcal{P}^n$.

(2.4) (Composition （合成）)

Let $n \in \mathbb{N}$ and $m \in \mathbb{N}^+$, let $\tau \in \mathcal{P}^{n+1}$, let $\psi \in \mathcal{P}^m$ and define $\phi(x_1, \ldots, x_n, y_1, \ldots, y_m) =_{df} \tau(x_1, \ldots, x_n, \psi(y_1, \ldots, y_m))$. Then $\phi \in \mathcal{P}^{n+m}$.

(2.5) (Primitive recursion (原始再帰))

Let $n \in \mathbb{N}^+$, let $\tau \in \mathcal{P}^n$ and let $\psi \in \mathcal{P}^{n+2}$. Then we have: if

$$\phi(x_1,\ldots,x_n,0) \quad =_{df} \tau(x_1,\ldots,x_n);$$
$$\phi(x_1,\ldots,x_n,y+1) =_{df} \psi(x_1,\ldots,x_n,y,\phi(x_1,\ldots,x_n,y)),$$

then $\phi \in \mathcal{P}^{n+1}$.

(2.6) (μ-recursion (μ-再帰)) Let $n \in \mathbb{N}^+$; then we have:

if $\tau \in \mathcal{P}^{n+1}$ and $\psi(x_1,\ldots,x_n) \ =_{df} \ \mu y[\tau(x_1,\ldots,x_n,y) = 1]$

$$=_{df} \begin{cases} \text{the smallest } y \text{ such that} \\ (1) \ \tau(x_1,\ldots,x_n,v) \quad \text{is defined for all } v \leq y; \\ (2) \ \tau(x_1,\ldots,x_n,v) \neq 1 \quad \text{for all } v < y; \\ (3) \ \tau(x_1,\ldots,x_n,y) = 1, \quad\quad\quad\quad\quad\quad \text{if such a } y \text{ exists;} \\ \text{not defined}, \quad\quad\quad\quad\quad\quad\quad\quad\quad\quad \text{otherwise;} \end{cases}$$

then $\psi \in \mathcal{P}^n$.

In the following we refer to the rules given above as operations. For example, when talking about primitive recursion, we refer to it as Operation (2.5). Note that all operations given above except Operation (2.5) are explicit. Operation (2.5) itself constitutes an *implicit* definition, since ϕ appears on both the left and right hand side. Thus, before we can continue, we need to verify whether or not Operation (2.5) does *always* define a function. This is by no means obvious. Recall that every implicit definition needs a justification. Therefore, we have to show the following theorem.

Theorem 11.2 (Dedekind's Justification Theorem) *If τ and ψ are functions, then there is precisely one function ϕ satisfying the scheme given in Operation (2.5).*

Proof. We have to show the uniqueness and existence of function ϕ.

11. Models of Computation

We start with the uniqueness.

Claim 1. There is at most one function ϕ satisfying the scheme given in Operation (2.5).

Suppose there are functions ϕ_1 and ϕ_2 satisfying the scheme given in Operation (2.5). We show by induction over y that

$$\phi_1(x_1,\ldots,x_n,y) = \phi_2(x_1,\ldots,x_n,y) \quad \text{for all } x_1,\ldots,x_n,y \in \mathbb{N}.$$

The induction basis is for $y = 0$. Then we directly get for all $x_1,\ldots,x_n \in \mathbb{N}$

$$\phi_1(x_1,\ldots,x_n,0) = \tau(x_1,\ldots,x_n) = \phi_2(x_1,\ldots,x_n,0).$$

As induction hypothesis (abbr. IH) we have that for all $x_1,\ldots,x_n \in \mathbb{N}$ and some $y \in \mathbb{N}$ it holds $\phi_1(x_1,\ldots,x_n,y) = \phi_2(x_1,\ldots,x_n,y)$.

The induction step is done from y to $y+1$. Using the scheme provided in Operation (2.5) we obtain

$$\begin{aligned}
\phi_1(x_1,\ldots,x_n,y+1) &= \psi(x_1,\ldots,x_n,y,\phi_1(x_1,\ldots,x_n,y)) &&\text{by definition}\\
&= \psi(x_1,\ldots,x_n,y,\phi_2(x_1,\ldots,x_n,y)) &&\text{by the IH}\\
&= \phi_2(x_1,\ldots,x_n,y+1) &&\text{by definition}.
\end{aligned}$$

Consequently $\phi_1 = \phi_2$, and Claim 1 is proved.

The following claim establishes the existence of function ϕ.

Claim 2. There exists a function ϕ satisfying the scheme given in Operation (2.5).

For showing the existence of ϕ we replace the inductive and implicit definition of ϕ by an infinite sequence of explicit definitions, i.e., let

$$\phi_0(x_1,\ldots,x_n,y) = \begin{cases} \tau(x_1,\ldots,x_n), & \text{if } y = 0; \\ \text{not defined}, & \text{otherwise}. \end{cases}$$

$$\varphi_1(x_1,\ldots,x_n,y) = \begin{cases} \varphi_0(x_1,\ldots,x_n,y), & \text{if } y < 1; \\ \psi(x_1,\ldots,x_n,0,\varphi_0(x_1,\ldots,x_n,0)), & \text{if } y = 1; \\ \text{not defined}, & \text{otherwise}. \end{cases}$$

$$\varphi_{i+1}(x_1,\ldots,x_n,y) = \begin{cases} \varphi_i(x_1,\ldots,x_n,y), & \text{if } y < i+1; \\ \psi(x_1,\ldots,x_n,i,\varphi_i(x_1,\ldots,x_n,i)), & \text{if } y = i+1; \\ \text{not defined}, & \text{otherwise}. \end{cases}$$

All definitions of the functions φ_i are *explicit*, and thus the functions φ_i exist by the set forming axiom. Consequently, for $y \in \mathbb{N}$ and every $x_1,\ldots,x_n \in \mathbb{N}$ the function φ defined by

$$\varphi(x_1,\ldots,x_n,y) =_{df} \varphi_y(x_1,\ldots,x_n,y)$$

does exist. Furthermore, by construction we directly get

$$\begin{aligned}
\varphi(x_1,\ldots,x_n,0) &= \varphi_0(x_1,\ldots,x_n,0) \\
&= \tau(x_1,\ldots,x_n) \text{ and}
\end{aligned}$$

$$\begin{aligned}
\varphi(x_1,\ldots,x_n,y+1) &= \varphi_{y+1}(x_1,\ldots,x_n,y+1) \\
&= \psi(x_1,\ldots,x_n,y,\varphi_y(x_1,\ldots,x_n,y)) \\
&= \psi(x_1,\ldots,x_n,y,\varphi(x_1,\ldots,x_n,y)),
\end{aligned}$$

and thus, φ is satisfying the scheme given in Operation (2.5). ■

Now, we are ready to define the class of all partial recursive functions.

11. Models of Computation

Definition 11.1 *We define the class $\mathcal{P}$ of all partial recursive functions (部分帰納的関数) to be the smallest function class containing the functions Z, S and V and all functions that can be obtained from Z, S and V by finitely many applications of the Operations* (2.1) *through* (2.6).

That is $\mathcal{P} = \bigcup_{n \in \mathbb{N}^+} \mathcal{P}^n$.

Furthermore, we define the important subclass of primitive recursive functions as follows.

Definition 11.2 *We define the class Prim of all primitive recursive functions (原始帰納的関数) to be the smallest function class containing the functions Z, S and V and all functions that can be obtained from Z, S and V by finitely many applications of the Operations* (2.1) *through* (2.5).

Note that, by definition, we have $Prim \subseteq \mathcal{P}$. We continue with some examples.

Example 11.1 *The identity function (恒等関数) $I: \mathbb{N} \to \mathbb{N}$ defined by $I(x) = x$ for all $x \in \mathbb{N}$ is primitive recursive.*

Proof. We want to apply Operation (2.4). Let $n = 0$ and $m = 1$. By our definition (cf. Step (1)), we know that $V, S \in \mathcal{P}^1$. So, V serves as the τ (note that $n + 1 = 0 + 1 = 1$) and S serves as the ψ in Operation (2.4) (note that $m = 1$). Consequently, the desired function I is the ϕ in Operation (2.4) (note that $n + m = 0 + 1 = 1$) and we can set

$$I(x) = V(S(x)) \ .$$

Hence, the identity function I is primitive recursive.

Example 11.2 *The binary addition function (二項和) $\alpha: \mathbb{N} \times \mathbb{N} \to \mathbb{N}$ given by $\alpha(n, m) = n + m$ for all $n, m \in \mathbb{N}$ is is primitive recursive.*

Proof. By assumption, $S \in \mathcal{P}$. As shown in Example 11.1, $I \in Prim$. First, we define some auxiliary functions by using the operations indicated below.

$\psi(x_1, x_2) \quad =_{df} S(x_1) \quad$ by using Operation (2.1);

$\tilde{\psi}(x_1, x_2) \quad =_{df} \psi(x_2, x_1) \quad$ by using Operation (2.3);

$\tau(x_1, x_2, x_3) =_{df} \tilde{\psi}(x_1, x_2) \quad$ by using Operation (2.1);

$\tilde{\tau}(x_1, x_2, x_3) =_{df} \tau(x_1, x_3, x_2) \quad$ by using Operation (2.3) .

Now, we are ready to apply Operation (2.5) for defining α, i.e., we set

$\alpha(n, 0) \quad =_{df} I(n),$

$\alpha(n, m+1) =_{df} \tilde{\tau}(n, m, \alpha(n, m))$.

Since we only used Operations (2.1) through (2.5), we see that $\alpha \in Prim$.

So, let us compute $\alpha(n, 1)$. Then we get

$\alpha(n, 1) = \alpha(n, 0+1) = \tilde{\tau}(n, 0, \alpha(n, 0))$

$= \tilde{\tau}(n, 0, I(n)) \quad$ by using $\alpha(n, 0) = I(n)$,

$= \tilde{\tau}(n, 0, n) \quad$ by using $I(n) = n$,

$= \tau(n, n, 0) \quad$ by using the definition of $\tilde{\tau}$,

$= \tilde{\psi}(n, n) \quad$ by using the definition of τ ,

$= \psi(n, n) \quad$ by using the definition of $\tilde{\psi}$,

$= S(n) = n+1 \quad$ by using the definition of ψ and S .

So, our definition may look more complex than necessary. In order to see, it is not, we compute $\alpha(n, 2)$.

$\alpha(n, 2) = \alpha(n, 1+1) = \tilde{\tau}(n, 0, \alpha(n, 1))$

$= \tilde{\tau}(n, 0, n+1) \quad$ by using $\alpha(n, 0) = n+1$,

$= \tau(n, n+1, 0)$

$= \tilde{\psi}(n, n+1)$

$= \psi(n+1, n)$

$= S(n+1) = n+2$.

11. Models of Computation

In the following we shall often omit some of the tedious technical steps. For example, in order to clarify that binary multiplication is primitive recursive, we simply point out that is suffices to set

$$m(x, 0) = Z(x),$$
$$m(x, y+1) = \alpha(x, m(x, y)).$$

Note that the constant 1 function c is primitive recursive, i.e., $c(n) = 1$ for all $n \in \mathbb{N}$. For seeing this, we set

$$c(0) = S(0),$$
$$c(n+1) = c(n).$$

In the following, instead of $c(n)$ we just write 1.

Now, we see that the signum function (符号関数) sg is in *Prim*, since

$$sg(0) = 0,$$
$$sg(n+1) = 1.$$

Since the natural numbers are not closed under subtraction, one conventionally uses the so-called arithmetic difference (数論的減算) defined as $m \dotminus n = m - n$ if $m \geq n$ and 0 otherwise. The arithmetic difference is primitive recursive, too, since for all $n, m \in \mathbb{N}$ we have

$$m \dotminus 0 = I(m),$$
$$m \dotminus (n+1) = V(m \dotminus n).$$

Occasionally, we shall also need $\overline{sg}$ defined as

$$\overline{sg}(n) = 1 \dotminus sg(n).$$

As the above definition shows, $\overline{sg}$ is also primitive recursive.

Moreover, we can easily extend binary addition and multiplication to any

11.1 Partial Recursive Functions

fixed number k of arguments. For example, in order to define the ternary addition function α^3 ($k = 3$), we apply Operation (2.4) and set

$$\alpha^3(x_1, x_2, x_3) =_{df} \alpha(x_1, \alpha(x_2, x_3)) \, .$$

Iterating this idea yields k-ary addition. Of course, we can apply it *mutatis mutandis* to multiplication. For the sake of notational convenience we shall use the more common $\sum_{i=1}^{k} x_i$ and $\prod_{i=1}^{k} x_i$ to denote k-ary addition and k-ary multiplication, respectively. Also, from now on we shall use the common $+$ and $\cdot$ to denote addition and multiplication whenever appropriate.

Example 11.3 *Let $g \in \mathcal{P}^{n+1}$ be any primitive recursive function. Then $f(x_1, \ldots, x_n, k) = \sum_{i=1}^{k} g(x_1, \ldots, x_n, i)$ is primitive recursive.*

Proof. By Example 11.2 we know that α is primitive recursive. Then the function f is obtained by applying Operations (2.5) and (2.4), i.e.,

$$f(x_1, \ldots, x_n, 0) = g(x_1, \ldots, x_n, 0) \, ,$$
$$f(x_1, \ldots, x_n, k+1) = \alpha(f(x_1, \ldots, x_n, k), g(x_1, \ldots, x_n, k+1)) \, .$$

Hence, f is primitive recursive. ∎

Analogously, one can show that the general multiplication is primitive recursive. That is, if g is as in Example 11.3 then $f(x_1, \ldots, x_n, k) = \prod_{i=1}^{k} g(x_1, \ldots, x_n, i)$ is primitive recursive.

Quite often one is defining functions by making case distinctions (cf., e.g., our definition of the predecessor function V). So, it is only natural to ask under what circumstances definitions by case distinctions do preserve primitive recursiveness. A convenient way to describe properties is the usage of *predicates* (述語). An n-ary predicate p over the natural numbers is a subset of N^n. Usually, one writes $p(x_1, \ldots, x_n)$ instead of

$(x_1, \ldots, x_n) \in p$. The characteristic function (特性関数) of an n-ary predicate p is the function $\chi_p \colon \mathbb{N}^n \to \{0, 1\}$ defined by

$$\chi_p(x_1, \ldots, x_n) = \begin{cases} 1, & \text{if } p(x_1, \ldots, x_n) \text{ ;} \\ 0, & \text{otherwise .} \end{cases}$$

A predicate p is said to be *primitive recursive* if χ_p is primitive recursive. Let p, q be n-ary predicates, then we define $p \wedge q$ to be the set $p \cap q$, $p \vee q$ to be the set $p \cup q$, and $\neg p$ to be the set $\mathbb{N}^n \setminus p$.

Lemma 11.1 *Let p, q be any primitive recursive n-ary predicates. Then $p \wedge q$, $p \vee q$, and $\neg p$ are also primitive recursive.*

Proof. Obviously, it holds

$$\chi_{p \wedge q}(x_1, \ldots, x_n) = \chi_p(x_1, \ldots, x_n) \cdot \chi_q(x_1, \ldots, x_n) ,$$
$$\chi_{p \vee q}(x_1, \ldots, x_n) = \chi_p(x_1, \ldots, x_n) + \chi_q(x_1, \ldots, x_n) ,$$
$$\chi_{\neg p}(x_1, \ldots, x_n) = 1 \dot{-} \chi_p(x_1, \ldots, x_n) .$$

Since we already know addition, multiplication and the arithmetic difference to be primitive recursive, the assertion of the lemma follows. ∎

Now, we can deal with functions defined by making case distinctions.

Theorem 11.3 *Let $p_1, \ldots, p_k$ be pairwise disjoint n-ary primitive recursive predicates, and let $\psi_1, \ldots, \psi_k \in \mathcal{P}^n$ be primitive recursive functions. Then the function $\gamma \colon \mathbb{N}^n \to \mathbb{N}$ defined by*

$$\gamma(x_1, \ldots, x_n) = \begin{cases} \psi_1(x_1, \ldots, x_n), & \text{if } p_1(x_1, \ldots, x_n) \text{ ;} \\ \vdots & \\ \psi_k(x_1, \ldots, x_n), & \text{if } p_k(x_1, \ldots, x_n) \text{ ;} \\ 0, & \text{otherwise ;} \end{cases}$$

is primitive recursive.

Proof. Since we can write γ as

$$\gamma(x_1,\ldots,x_n) = \sum_{i=1}^{k} \chi_{P_i}(x_1,\ldots,x_n) \cdot \psi_i(x_1,\ldots,x_n),$$

the theorem follows from the primitive recursiveness of general addition and multiplication. ∎

11.2 Pairing Functions

Quite often it would be very useful to have a bijection from $\mathbb{N} \times \mathbb{N}$ to $\mathbb{N}$. So, first we have to ask whether or not such a bijection does exist. This is indeed the case. Recall that the elements of $\mathbb{N} \times \mathbb{N}$ are ordered pairs of natural numbers. So, we may easily represent all elements of $\mathbb{N} \times \mathbb{N}$ in a two dimensional array, where row x contains all pairs (x, y), i.e., having x in the first component and $y = 0, 1, 2, \ldots$ (cf. Figure 11.1).

(0,0)	(0,1)	(0,2)	(0,3)	(0,4)	...
(1,0)	(1,1)	(1,2)	(1,3)	(1,4)	...
(2,0)	(2,1)	(2,2)	(2,3)	(2,4)	...
(3,0)	(3,1)	(3,2)	(3,3)	(3,4)	...
(4,0)	(4,1)	(4,2)	(4,3)	(4,4)	...
(5,0)	...				
...	...				

Figure 11.1 A two dimensional array representing $\mathbb{N} \times \mathbb{N}$.

The resulting bijection is shown in Figure 11.2. That is, we arrange all these pairs in a sequence starting

$$(0,0), (0,1), (1,0), (0,2), (1,1), (2,0), (0,3), (1,2), \ldots . \quad (11.1)$$

In this order, all pairs (x, y) appear before all pairs (x', y') if and only if $x + y < x' + y'$. So they are arranged in order of incrementally growing component sums. The pairs with the same component sum are ordered by

x\y	0	1	2	3	4	5 ...
0	0	1	3	6	10	✓
1	2	4	7	11	✓	
2	5	8	12	✓		
3	9	13	✓			
4	14	✓				
5	✓					
⋮						

Figure 11.2 The bijection.

the first component starting with the smallest one. That is, pair (x, y) is located in the segment

$$(0, x+y), \ (1, x+y-1), \ \ldots, \ (x, y), \ \ldots, \ (x+y, 0) \ . \qquad (11.2)$$

Note that there are $x+y+1$ many pairs having the component sum $x+y$. Thus, in front of pair $(0, x+y)$ in the Sequence (11.1) we have $x+y$ many segments containing a total of $1 + 2 + 3 + \cdots + (x+y)$ many pairs.

Taking into account that

$$\sum_{i=0}^{n} i = \frac{n(n+1)}{2} = \sum_{i=1}^{n} i \,, \qquad (11.3)$$

we thus can define the desired bijection $c \colon \mathbb{N} \times \mathbb{N} \to \mathbb{N}$ by setting

$$c(x, y) = x + \sum_{i=1}^{x+y} i = x + \frac{(x+y)(x+y+1)}{2} \qquad \text{by (11.3)}$$

$$= \frac{(x+y)^2 + 3x + y}{2} \,. \qquad (11.4)$$

Note that we start counting with 0 in the Sequence (11.1), since otherwise we would not obtain a bijection (cf. Figure 11.2). So, pair $(0, 0)$ is at the 0th position, pair $(0, 1)$ at the 1st, ... Let us make a quick check by computing $c(1, 0)$ and $c(0, 2)$. We directly obtain $c(1, 0) = ((1 + 0)^2 + 3 + 0)/2 = 2$ and $c(0, 2) = ((0 + 2)^2 + 2)/2 = 3$.

The unary function $f(n) = \sum_{i=0}^{n} i$ is primitive recursive (cf. Example 11.3). Thus, c is obtained from f by applying Operation (2.4) and then again binary addition (i.e., the function α defined in Example 11.2). Thus we can conclude that c is primitive recursive, too.

Note that c is referred to as *Cantor's pairing function* (カントールの対関数). For the two inverse functions d_1 and d_2 such that for all $x, y \in \mathbb{N}$, if $z = c(x, y)$ then $x = d_1(z)$ and $y = d_2(z)$, we refer to Problem 11.2.

Exercise 11.1 *Show that for every fixed* $k \in \mathbb{N}$, $k > 2$, *there is a primitive recursive bijection* $c_k \colon \mathbb{N}^k \to \mathbb{N}$.

Exercise 11.2 *Let* $\mathbb{N}^*$ *be the set of all finite sequences of natural numbers. Show that there is a primitive recursive bijection* $c_* \colon \mathbb{N}^* \to \mathbb{N}$.

Exercise 11.3 *Prove that*

(1) *every constant function,*

(2) $f(n) = 2^n$, *and* $d(n) = 2^{2^n}$,

(3) min *and* max *of two or more arguments,*

(4) $|m - n|$,

(5) $\left\lfloor \dfrac{n}{m} \right\rfloor$ (*i.e., division without remainder*) *and the remainder of division are primitive recursive.*

11.3 General Recursive Functions

Next, we define the class of *general recursive functions* (一般帰納的関数).

Definition 11.3 *For all* $n \in \mathbb{N}^+$ *we define* $\mathcal{R}^n$ *to be the set of all functions* $f \in \mathcal{P}^n$ *such that* $f(x_1, \ldots, x_n)$ *is defined for all* $x_1, \ldots, x_n \in \mathbb{N}$. *Furthermore, we set* $\mathcal{R} = \bigcup_{n \in \mathbb{N}^+} \mathcal{R}^n$.

In other words, $\mathcal{R}$ is the set of all functions that are total and partial recursive. Now, we can show the following theorem.

Theorem 11.4 $Prim \subset \mathcal{R} \subset \mathcal{P}$.

Proof. Clearly $Z, S, V \in \mathcal{R}$. Furthermore, after a bit of reflection it should be obvious that any finite number of applications of Operations (2.1) through (2.5) results only in total functions. Hence, every primitive recursive function is general recursive, too. This shows $Prim \subseteq \mathcal{R}$. Also, $\mathcal{R} \subseteq \mathcal{P}$ is obvious by definition. So, it remains to show that the two inclusions are proper. This is done by the following claims.

Claim 1. $\mathcal{P} \setminus \mathcal{R} \neq \emptyset$.

By definition, $S \in \mathcal{P}$. Furthermore, using Operation (2.4), it is easy to see that $\delta(n) =_{df} S(S(n))$ is in $\mathcal{P}$, too. Now, note that $\delta(n) = n + 2 > 1$ for all $n \in \mathbb{N}$.

Using Operation (2.1) we define $\tau(x, y) = \delta(y)$, and therefore $\tau \in \mathcal{P}$. Consequently,

$$\psi(x) = \mu[\tau(x, y) = 1] \qquad (11.5)$$

is the *nowhere defined function*, and hence $\psi \notin \mathcal{R}$. On the other hand, by construction $\psi \in \mathcal{P}$. Therefore, we get $\psi \in \mathcal{P} \setminus \mathcal{R}$, and Claim 1 is shown.

Claim 2. $\mathcal{R} \setminus Prim \neq \emptyset$.

Showing this claim is much more complicated. First, we define a function

$$\text{ap}(0, m) = m + 1 , \qquad (11.6)$$

$$\text{ap}(n + 1, 0) = \text{ap}(n, 1) , \qquad (11.7)$$

$$\text{ap}(n + 1, m + 1) = \text{ap}(n, \text{ap}(n + 1, m)) . \qquad (11.8)$$

The function ap is the so-called Ackermann-Péter function (アッカーマン-ペータ関数). Hilbert conjectured in 1926 that every total and computable function is also primitive recursive. This conjecture was disproved by Ackermann [1] in 1928 and Péter [27] simplified Ackermann's definition in 1955.

11.3 General Recursive Functions

Now, it suffices to show that function ap is *not* primitive recursive and that function ap is general recursive. Both parts are not easy to prove. So, due to the lack of space, we must skip some parts. But before we start, let us confine ourselves that the function ap is *intuitively* computable. For doing this, consider the following fragment of pseudo-code implementing the function ap as peter.

```
function peter(n, m)
    if n = 0
        return m + 1
    else if m = 0
        return peter(n - 1, 1)
    else
        return peter(n - 1, peter(n, m - 1))
```

Next, we sketch the proof that the function ap cannot be primitive recursive. First, for every primitive recursive function ϕ, one defines a function f_ϕ as follows. Let k be the arity of ϕ; then we set

$$f_\phi(n) = \max\left\{\phi(x_1,\ldots,x_k) \mid \sum_{i=1}^{k} x_i \leq n\right\}. \tag{11.9}$$

Then, by using the inductive construction of the class *Prim*, one can show by structural induction that for every primitive recursive function ϕ there is a number $n_\phi \in \mathbb{N}$ such that

$$f_\phi(n) < \mathrm{ap}(n_\phi, n) \quad \text{for all } n \geq n_\phi. \tag{11.10}$$

Intuitively, the latter statement shows that the Ackermann-Péter function grows faster than every primitive recursive function.

The rest is easy. Suppose ap $\in$ *Prim*. Then, taking into account that the identity function I is primitive recursive (cf. Example 11.1), one directly sees by application of Operation (2.4) that

$$\kappa(n) = ap(I(n), I(n)) \tag{11.11}$$

is primitive recursive, too. By (11.10), for κ there exists a number $n_\kappa \in \mathbb{N}$ such that

$$f_\kappa(n) < ap(n_\kappa, n) \quad \text{for all } n \geq n_\kappa . \tag{11.12}$$

But now, putting (11.9), (11.12) and (11.11) together, we obtain

$$\kappa(n_\kappa) \leq f_\kappa(n_\kappa) < ap(n_\kappa, n_\kappa) = \kappa(n_\kappa) ,$$

a contradiction.

For the second part, one has to prove that $ap \in \mathcal{R}$ which mainly means to provide a construction to express the function ap using the Operations (2.1) through (2.5) *and* the μ-operator. We refer the interested reader to Hermes [14]. ∎

11.4 Problem Set 11

Problem 11.1 Show the binary predicates $=, <,$ and $\leq$ defined as usual over $\mathbb{N} \times \mathbb{N}$ to be primitive recursive.

Problem 11.2 Determine the functions d_1 and d_2 such that for all $x, y \in \mathbb{N}$, if $z = c(x, y)$ then $x = d_1(z)$ and $y = d_2(z)$. Prove or disprove $d_1, d_2 \in \mathit{Prim}$.

Problem 11.3 Try to compute the following values of the Ackermann-Péter function: $ap(1, m), ap(2, m), ap(3, m),$ and $ap(4, m)$ for $m = 0, 1, 2$.

12
Turing Machines

After having dealt with partial recursive functions, we turn our attention to Turing machines (チューリング機械) introduced by Alan Turing [38]. We use TM as abbreviation for Turing machine. His idea was to formalize the notion of "intuitively computable" functions by using the four properties of an algorithm given at the beginning of Chapter 11. Starting from these properties, he observed that the primitive operations could be reduced to a level such that a machine can execute the whole algorithm. For the sake of simplicity, here we consider one-tape TMs (1 テープ チューリング機械).

12.1 One-tape Turing Machines

A one-tape TM consists of an infinite tape which is divided into cells. Each cell can contain exactly one of the tape-symbols. Initially, we assume that all cells of the tape contain the symbol $*$ except those in which the actual input has been written. Moreover, we enumerate the borders between the tape cells as shown in Figure 12.1. We use these numbers to refer to the cell right to the border. The TM does neither know nor use these numbers.

Furthermore, the TM possesses a read-write head. This head can observe one cell at a time. Additionally, the machine has a finite number of states it can be in and a set of instructions it can execute. Initially, it is always

12. Turing Machines

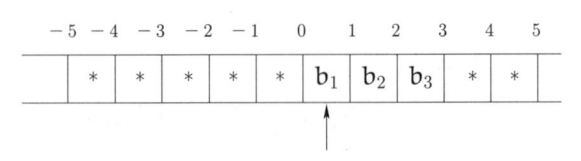

	−5	−4	−3	−2	−1	0	1	2	3	4	5
	*	*	*	*	*	b_1	b_2	b_3	*	*	

Figure 12.1 The tape of a TM with input $b_1 b_2 b_3$.

in the start state z_s and the head is observing the leftmost symbol of the input, i.e., the cell 0. We indicate the position of the head by an arrow pointing to it (cf. Figure 12.1).

Then, the TM works as follows. When in state z and reading tape symbol b, it writes tape symbol b' into the observed cell, changes its state to z' and moves the head either to the left (denoted by L) or to the right (denoted by R) or does not move the head (denoted by N) provided (z, b, b', H, z') is in the instruction set of the Turing machine, where $H \in \{L, N, R\}$. The execution of one instruction is called *step*. When the machine reaches the *distinguished* state z_f (the *final* state), it stops. Thus, formally, we can define a Turing machine as follows.

Definition 12.1 *A triple* $M = [B, Z, A]$ *is called* deterministic one-tape TM (決定性 1 テープ チューリング機械) *if* B, Z, A *are non-empty finite sets such that* $B \cap Z = \emptyset$ *and*

(1) $\operatorname{card}(B) \geq 2$ ($B = \{*, |, \ldots\}$) (*tape-symbols* (テープ記号)),

(2) $\operatorname{card}(Z) \geq 2$ ($Z = \{z_s, z_f, \ldots\}$) (*set of states* (状態集合)),

(3) $A \subseteq Z \setminus \{z_f\} \times B \times B \times \{L, N, R\} \times Z$ (*instruction set* (命令集合)), *where for every* $z \in Z \setminus \{z_f\}$ *and every* $b \in B$ *there is precisely one 5-tuple* $(z, b, \cdot, \cdot, \cdot)$.

Often, we represent the instruction set A in a table (cf. Figure 12.2).

If the instruction set is small, it often convenient to write $zb \to b'Hz'$, where $H \in \{L, N, R\}$ instead of (z, b, b', H, z'). Also, we usually refer to the instruction set of a TM M as to the *program* (プログラム) of M.

	*	|	b_2	...	b_n
z_s	$b'Nz_3$				
z_1	.				
.	.				
.	.				
z_n	.				

Figure 12.2 A Turing table.

12.2 Turing Computations

Next, we explain how a TM is computing a function. Our primary concern are functions from $\mathbb{N}^n$ to $\mathbb{N}$, i.e., $f: \mathbb{N}^n \to \mathbb{N}$. Therefore, the inputs are tuples $(x_1, \ldots, x_n) \in \mathbb{N}^n$. We shall reserve the special tape symbol $\#$ to separate x_i from x_{i+1}. Moreover, for the sake of simplicity, in the following we shall assume that numbers are unary encoded (1 進符号化), e.g., number 0 is represented by $*$, number 1 by $|$, number 2 by $\|$, number 3 by $\|\|$, a.s.o. Note that this convention is no restriction as long as we do not consider the *complexity* (計算量) of a Turing computation.

We introduce two notations. Let $f: \mathbb{N}^n \to \mathbb{N}$ be any function. If the value $f(x_1, \ldots, x_n)$ is not defined for a tuple $(x_1, \ldots, x_n) \in \mathbb{N}^n$ then we write $f(x_1, \ldots, x_n) \uparrow$. If $f(x_1, \ldots, x_n)$ is defined then we write $f(x_1, \ldots, x_n) \downarrow$.

Definition 12.2 *Let* M *be any TM, let* $n \in \mathbb{N}^+$ *and let* $f: \mathbb{N}^n \to \mathbb{N}$ *be any function. We say that* M *computes the function* f (M が関数 f を計算する) *if for all* $(x_1, \ldots, x_n) \in \mathbb{N}^n$ *the following conditions are satisfied:*

(1) *If* $f(x_1, \ldots, x_n) \downarrow$ *and if* $x_1 \# \ldots \# x_n$ *is written on the empty tape of* M (*beginning in cell* 0) *and* M *is started on the leftmost symbol of* $x_1 \# \ldots \# x_n$ *in state* z_s, *then* M *stops after having executed finitely many steps in state* z_f. *Moreover, if* $f(x_1, \ldots, x_n) = 0$, *then the symbol observed by* M *in state* z_f *is* $*$. *If* $f(x_1, \ldots, x_n) \neq 0$, *then the string*

beginning in the cell observed by M in state z_f (read from left to right) of consecutive | denotes the results (cf. Figure 12.3).

(2) If $f(x_1,\ldots,x_n) \uparrow$ and if $x_1\#\ldots\#x_n$ is written on the empty tape of M (beginning in cell 0) and M is started on the leftmost symbol of $x_1\#\ldots\#x_n$ in state z_s then M does not stop.

By f_M^n we denote the function from $\mathbb{N}^n$ to $\mathbb{N}$ computed by M.

	−5	−4	−3	−2	−1	0	1	2	3	4	5
	*	*	*	*	*	|	|	|		*	*

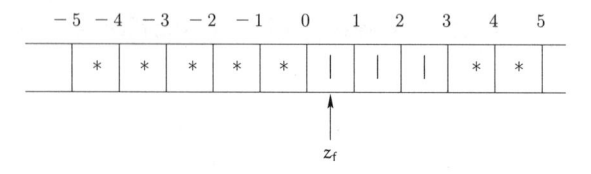

z_f

Figure 12.3 The tape of a TM with result 3 (written as |||).

Definition 12.3 *Let* $n \in \mathbb{N}^+$ *and let* $f: \mathbb{N}^n \to \mathbb{N}$ *be any function. The function* f *is said to be* Turing computable (チューリング計算可能) *if there exists a Turing machine* M *such that* $f_M^n = f$. *We define*

$\mathcal{T}^n =$ the set of all n-ary Turing computable functions.

$\mathcal{T} = \bigcup_{n \in \mathbb{N}^+} \mathcal{T}^n =$ the set of all Turing computable functions.

Furthermore, as usual we use $dom(f)$ to denote the domain (定義域) of function f, and $range(f)$ to denote the range (値域) of function f.

Now, it is only natural to ask which functions are Turing computable. The answer is provided by the following theorem.

Theorem 12.1 *The class of Turing computable functions is equal to the class of partial recursive functions, i.e.,* $\mathcal{T} = \mathcal{P}$.

Proof. For showing $\mathcal{P} \subseteq \mathcal{T}$ it suffices to prove the following:

(1) The functions Z, S and V are Turing computable.

(2) The class of Turing computable functions is closed under the Operations (2.1) through (2.6) defined in Chapter 11.

First, we define a TM M computing the constant zero function Z. Let $M = [\{*, |\}, \{z_s, z_f\}, A]$, where the set A contains the following instructions:

$$z_s \,|\, \to\, |\,L z_f\,, \quad \text{and} \quad z_s * \,\to\, * N z_f\,.$$

That is, if the input is not zero, then M moves its head one position to the left and stops. By our definition of a TM, then M observes in cell -1 a $*$, and thus its output is 0. If the input is zero, then M observes in cell 0 a $*$, leaves it unchanged, does not move its head and stops. Clearly, $f_M^1(x) = 0$ for all $n \in \mathbb{N}$.

The successor function S is computed by the following TM M. We set $M = [\{*, |\}, \{z_s, z_f\}, A]$, where the set A contains the following instructions:

$$z_s \,|\, \to\, |\,L z_s\,, \quad \text{and} \quad z_s * \,\to\, |\,N z_f\,.$$

That is, the TM M just adds a $|$ to its input and stops. Thus, we have $f_M^1(x) = S(x)$ for all $n \in \mathbb{N}$.

The predecessor function V is computed by $M = [\{*, |\}, \{z_s, z_f\}, A]$, where A is the following set of instructions:

$$z_s \,|\, \to\, * R z_f\,, \quad \text{and} \quad z_s * \,\to\, * N z_f\,.$$

Now, the TM either observes a $|$ in cell 0 which it removes and then the head goes one cell to the right or it observes a $*$, and stops without moving its head. Hence, $f_M^1(x) = V(x)$ for all $n \in \mathbb{N}$. This proves Part (1).

Next, we sketch the proof of Part (2). This is done in a series of claims.

Claim 1. (Introduction of fictitious variables) *Let $n \in \mathbb{N}^+$; then we have: if $\tau \in \mathfrak{T}^n$ and $\psi(x_1, \ldots, x_n, x_{n+1}) = \tau(x_1, \ldots, x_n)$, then $\psi \in \mathfrak{T}^{n+1}$.*

Intuitively, it is clear that Claim 1 holds. In order to compute ψ, all we have to do is to remove x_{n+1} from the input tape, then moving the head back into the initial position and then we start the Turing program for τ. Consequently, $\psi \in \mathfrak{T}^{n+1}$. We omit the details.

Claim 2. (Identifying variables) *Let $n \in \mathbb{N}^+$; then we have:*
if $\tau \in \mathcal{T}^{n+1}$ and $\psi(x_1, \ldots, x_n) = \tau(x_1, \ldots, x_n, x_n)$, then $\psi \in \mathcal{T}^n$.

For proving Claim 2, we only need a Turing program that copies the last variable (that is, x_n). Thus, the initial tape inscription $**x_1\#\ldots\#x_n**$ is transformed into $**x_1\#\ldots\#x_n\#x_n**$ and the head is moved back into its initial position and M is put into the initial state of the program computing τ. Now, we start the program for τ. Consequently, $\psi \in \mathcal{T}^n$. Again, we omit the details.

Claim 3. (Permuting variables)
Let $n \in \mathbb{N}^+$, $n \geq 2$ and let $i \in \{1, \ldots, n\}$; then we have: if $\tau \in \mathcal{T}^n$ and $\psi(x_1, \ldots, x_i, x_{i+1}, \ldots, x_n) = \tau(x_1, \ldots, x_{i+1}, x_i, \ldots, x_n)$, then $\psi \in \mathcal{T}^n$.

Claim 3 can be shown *mutatis mutandis* as Claim 2, and we therefore omit its proof here.

Claim 4. (Composition) *Let $n \in \mathbb{N}$ and $m \in \mathbb{N}^+$. Let $\tau \in \mathcal{T}^{n+1}$, let $\psi \in \mathcal{T}^m$ and let $\phi(x_1, \ldots, x_n, y_1, \ldots, y_m) = \tau(x_1, \ldots, x_n, \psi(y_1, \ldots, y_m))$. Then $\phi \in \mathcal{T}^{n+m}$.*

The proof of Claim 4 is more complicated. Clearly, the idea is to move the head to the right until it observes the first symbol of y_1. Then we could start the Turing program for ψ. If $\psi(y_1, \ldots, y_m) \uparrow$, then the machine for ϕ also diverges on input $x_1, \ldots, x_n, y_1, \ldots, y_m$. But if $\psi(y_1, \ldots, y_m) \downarrow$, our goal would be to obtain the new tape inscription

$$**x_1\#\ldots\#x_n\#\psi(y_1, \ldots, y_m)**$$

and then to move the head to the left such that it observes the first symbol of x_1. This would allow us to start then the Turing program for τ. So, the difficulty we have to overcome is to ensure that the computation of $\psi(y_1, \ldots, y_m)$ does *not overwrite* the x_i. Therefore, we need the following lemma.

12.2 Turing Computations

Lemma M^+. *For every TM M there exists a TM M^+ such that*

(1) $f_M^n = f_{M^+}^n$ *for all n.*

(2) M^+ *is never moving left to the initial cell observed when starting its computation.*

(3) *For all $x_1, \ldots, x_n$: If $f_{M^+}^n(x_1, \ldots, x_n) \downarrow$, then the computation stops with the head observing the same cell it has observed when the computation started and right to the result computed there are only $*$ on the tape.*

We show Lemma M^+ as follows. The TM M^+ does the following.

- It moves the whole input one cell to the right,
- it marks the initial cell with a special symbol, say L,
- it marks the first cell right to the moved input with a special symbol, say E,
- it works then as M does except the following three exceptions:
  - If M^+ reaches the cell marked with E, say in state z, then it moves the marker E one cell to the right, moves the head then one position to the left and writes a $*$ into this cell (that is into the cell that originally contained E) and continues to simulate M in state z with the head at this position (that is, the cell originally containing E and now containing $*$).
  - If M in state z enters the cell containing L, then the whole tape inscription between L and E (including E but excluding L) is moved one position to the right. In the cell rightmost to L a $*$ is written and M^+ continues to simulate M observing this cell.
  - If M stops, then M^+ moves the whole result left such that the first symbol of the result is now located in the cell originally

containing L. Furthermore, into all cells starting from the first position right to the moved result and ending in E a $*$ is written and then the head is moved back to the leftmost cell containing the result. Then M^+ also stops.

This proves Lemma M^+.

Having Lemma M^+, now Claim 4 follows as described above.

The remaining claims for primitive recursion and μ-recursion are left as an exercise (see Problems 12.2 and 12.3). This shows $\mathcal{P} \subseteq \mathcal{T}$.

Finally, we have to show $\mathcal{T} \subseteq \mathcal{P}$. Let $n \in \mathbb{N}^+$, let $f \in \mathcal{T}^n$ and let M be any Turing machine computing f. We define the functions t (time) and r (result) as follows.

$$t(x_1, \ldots, x_n, y) = \begin{cases} 1, & \text{if } M, \text{ when started on } x_1, \ldots, x_n, \\ & \text{stops after having executed at most } y \text{ steps;} \\ 0, & \text{otherwise.} \end{cases}$$

$$r(x_1, \ldots, x_n, y) = \begin{cases} f(x_1, \ldots, x_n), & \text{if } t(x_1, \ldots, x_n, y) = 1; \\ 0, & \text{otherwise.} \end{cases}$$

Now, one can show that $t, r \in Prim$. Furthermore, Kleene [19] showed the following normal form theorem using t and r as defined above.

Theorem 12.2 *For every function* $f \in \mathcal{T}^n$, $n \in \mathbb{N}^+$ *there are functions* $t, r \in Prim$ *such that*

$$f(x_1, \ldots, x_n) = r(x_1, \ldots, x_n, \mu y[t(x_1, \ldots, x_n, y) = 1]) \quad (12.1)$$

for all $x_1, \ldots, x_n \in \mathbb{N}^n$.

We do not prove this theorem here, since a proof is beyond the scope of this course.

Assuming Theorem 12.2, the inclusion $\mathcal{T} \subseteq \mathcal{P}$ follows from the primitive-recursiveness of t and r and Equation (12.1), since the latter one shows

12.2 Turing Computations

that one has to apply the μ-operator exactly ones (Operation (2.6)) and the resulting function is composed with r by using Operation (2.4). Consequently, $f \in \mathcal{P}$ and the theorem follows. ∎

Theorem 12.2 is of fundamental epistemological importance. Though we started from completely different perspectives, we finally arrived at the same set of computable functions. Subsequently, different approaches have been proposed to formalize the notion of "intuitively computable" functions. These approaches comprise, among others, Post algorithms [28], Markov algorithms [23], Random-access machines (ランダムアクセス機械) [7], [13], [34], and Church's [5] λ-calculus (λ-計算).

As it turned out, all these formalizations define the same set of computable functions, i.e., the resulting class of functions is equal to the Turing computable functions. This led Church to his famous thesis.

Church's Thesis (チャーチの提唱) *The class of the "intuitively computable" functions is equal to the class of Turing computable functions.*

Note that Church's Thesis can neither be proved nor disproved in a formal way, since it contains the not defined term "intuitively computable." Whenever this term is defined in a mathematical precise way, one gets a precisely defined class of functions for which one then could try to prove a theorem as we did above.

It should be noted that the notion of a Turing machine contains some idealization such as potentially unlimited time and and having access to unlimited quantities of cells on the Turing tape. On the one hand, this idealization will limit the usefulness of theorems showing that some function is Turing computable. On the other hand, if one can prove that a particular function is *not* Turing computable, then such a result is very strong.

12.3 The Universal Turing Machine

In this section we are going to show that there is *one* Turing machine which can compute all partial recursive functions.

First, using our results concerning pairing functions, it is easy to see that we can encode any n-tuple of natural numbers into a natural number. As we have seen, this encoding is primitive recursive. Thus, in the following, we use $\mathcal{P}$ to denote the set of all partial recursive functions from $\mathbb{N}$ to $\mathbb{N}$.

Next, we consider any partial recursive function $\psi(i, x)$, i.e., $\psi: \mathbb{N}^2 \to \mathbb{N}$. Thus, if we fix the first argument i, then we obtain a partial recursive function of one argument. Usually, one uses the notation from the λ calculus to specify the argument which is not fixed, i.e., we write $\lambda x.\psi(i, x)$ to denote the partial recursive function of the argument x. It is also common to write just ψ_i instead of $\lambda x.\psi(i, x)$. Thus, we can visualize all functions of one argument computed by ψ as shown in Figure 12.4.

ψ_0	$\psi(0,0)$	$\psi(0,1)$	$\psi(0,2)$	$\psi(0,3)$	$\psi(0,4)$	...
ψ_1	$\psi(1,0)$	$\psi(1,1)$	$\psi(1,2)$	$\psi(1,3)$	$\psi(1,4)$	...
ψ_2	$\psi(2,0)$	$\psi(2,1)$	$\psi(2,2)$	$\psi(2,3)$	$\psi(2,4)$	...
ψ_3	$\psi(3,0)$	$\psi(3,1)$	$\psi(3,2)$	$\psi(3,3)$	$\psi(3,4)$	...
ψ_4	$\psi(4,0)$	$\psi(4,1)$	$\psi(4,2)$	$\psi(4,3)$	$\psi(4,4)$	...
ψ_5	$\psi(5,0)$	...				
.						
.						
.						
ψ_i	...	...				
...						

Figure 12.4 A two dimensional array representing all ψ_i.

For having an example, consider $\psi(i, x) = ix$; then e.g., $\psi_7(x) = 7x$.

So it is justified to call every function $\psi \in \mathcal{P}^2$ a *numbering* (番号付け).

12.3 The Universal Turing Machine

Definition 12.4 *A numbering* $\psi \in \mathcal{P}^2$ *is said to be* universal (万能) *for $\mathcal{P}$ if* $\{\psi_i \mid i \in \mathbb{N}\} = \mathcal{P}$.

Clearly, now the interesting question is whether or not a universal $\psi \in \mathcal{P}^2$ for $\mathcal{P}$ does exist. If there is a universal ψ for $\mathcal{P}$, then, by Theorem 12.1, we know that ψ is Turing computable, too. Therefore, we could interpret any TM computing ψ as a *universal Turing machine* (万能チューリング機械). The following theorem establishes the existence of a universal ψ.

Theorem 12.3 *There exists a universal numbering* $\psi \in \mathcal{P}^2$ *for $\mathcal{P}$.*

Proof. (Sketch) The idea is easily explained. By Theorem 12.1 we know that for every $\tau \in \mathcal{P}$ there is TM M such that $f_M^1 = \tau$. Therefore, we aim to encode every TM into a natural number. Thus, we need an injective partial recursive function *cod* such that $cod(M) \in \mathbb{N}$. In order to make this idea work we also need a general recursive function *decod* such that

$$decod(cod(M)) = \text{program of M}.$$

If the input i to *decod* is not a correct encoding of some TM, then we set $decod(i) = 0$.

The universal function ψ is then described by a TM U taking two arguments as input, i.e., i and x. When started as usual, U first computes $decod(i)$. If $decod(i) = 0$, then U computes the function Z. Otherwise, it should simulate the program of the machine M returned by $decod(i)$.

To realize this behavior, the following additional conditions must be met:

(1) U is not allowed to overwrite the program obtained from the computation of $decod(i)$,

(2) U must be realized by using only finitely many tape symbols and a finite set of states.

Next, we shortly explain how all our conditions can be realized. For the sake of better readability, in the following we always denote the tape symbols by b_i and the state sets always starts with $z_s, z_f, \ldots$. Let

$$M = [\{b_1, \ldots, b_m\}, \{z_s, z_f, z_1, \ldots, z_k\}, A]$$

be given. Then we use the following coding (here we write 0^n to denote the string consisting of exactly n zeros):

$$cod(L) = 101$$
$$cod(R) = 1001$$
$$cod(N) = 10001$$
$$cod(z_s) = 10^4 1$$
$$cod(z_f) = 10^6 1$$
$$cod(z_\ell) = 10^{2(\ell+3)} 1 \qquad \text{for all } \ell \in \{1, \ldots, k\} ,$$
$$cod(b_\ell) = 10^{2(\ell+1)+1} 1 \qquad \text{for all } \ell \in \{1, \ldots, m\} .$$

The instruction set is then encoded by concatenating the codings of its parts, that is

$$cod(zb \to b'Hz') = cod(z) cod(b) cod(b') cod(H) cod(z') .$$

For example, $z_s b_1 \to b_2 N z_1$ is then encoded as

$$10^4 1 10^5 1 10^7 1 10001 10^8 1 .$$

We have m tape symbols and $k+2$ states. Thus, there must be $m(k+1)$ many instructions $I_1, \ldots, I_{m(k+1)}$ (cf. Definition 12.1) which we assume to be written down in canonical order. Consequently, we finally encode the program of M by concatenating the encodings of all these instructions, i.e.,

$$cod(M) = cod(I_1) \cdots cod(I_{m(k+1)}) .$$

This string is interpreted as a natural number written in binary.

12.3 The Universal Turing Machine

Now, it is easy to see that *cod* is computable and that it is injective, i.e., if $M \neq M'$ then $cod(M) \neq cod(M')$.

Furthermore, if we use the function *cod* as described above, then *decode* reduces to check algorithmically that an admissible string is given. This is easy. If it is, the program of M can be directly read from the string.

Finally, we have to describe how the simulation is done. First, we have to ensure that U is not destroying the program of M. This is essentially done as outlined in Lemma M^+. Thus, it remains to explain how Condition (2) is realized. Clearly, U cannot memorize the actual state of M during simulation in its state set, since this would potentially require an unlimited number of states. But U can mark the actual state in which M is on its tape (e.g., by using bold letters).

In order to ensure that the TM U is only using finitely many tape symbols, U is not using directly b_ℓ from M's tape alphabet but just $cod(b_\ell) = 10^{2(\ell+1)+1}1$. This requires just two tape symbols for the simulation. We omit the details.

The Turing machine U can thus be expressed as a partial recursive function $\psi \in \mathcal{P}^2$ via Theorem 12.1. ∎

Summarizing, we have constructed a Turing machine U that can simulate every TM computing a function of one argument. Since we can encode any tuple of natural numbers into a natural number, we thus have a *universal Turing machine*.

Corollary 12.1 *There exists a universal Turing machine* U *for* $\mathcal{T}$.

We finish this chapter by shortly explaining how Turing machines can accept formal languages.

12.4 Accepting Languages

Let Σ denote any finite alphabet. Again, we use Σ^* to denote the free monoid over Σ and λ to denote the empty string. Note that $\lambda \neq *$.

Next, we define what does it mean that a TM is accepting a language L.

Definition 12.5 *A language* $L \subseteq \Sigma^*$ *is* accepted by TM M *(TM M に より受理される) if for every* $w \in \Sigma^*$ *the following conditions are satisfied.*

If w is written on the empty tape of M (beginning in cell 0) and the TM M is started on the leftmost symbol of w in state z_s then M stops after having executed finitely many steps in state z_f. Moreover,

(1) *if* $w \in L$ *then the cell observed by M in state z_f contains a* $|$.

In this case we also write $M(w) = |$.

(2) *If* $w \notin L$ *then the cell observed by M in state z_f contains a* $*$.

In this case we also write $M(w) = *$.

Of course, in order to accept a language $L \subseteq \Sigma^*$ by a Turing machine $M = [B, Z, A]$ we always have to assume that $\Sigma \subseteq B$.

Moreover, for every Turing machine M we define

$$L(M) = \{w \mid w \in \Sigma^* \text{ and } M(w) = |\}, \qquad (12.2)$$

and we refer to $L(M)$ as to the language accepted by M.

Example 12.1 Let $\Sigma = \{a\}$ and $L = \Sigma^+$.

We set $B = \{*, a, |\}$, $Z = \{z_s, z_f\}$ and define A as follows.

$$z_s * \to * N z_f, \quad z_s a \to | N z_f, \quad \text{and} \quad z_s | \to | N z_s.$$

Note that we have included the instruction $z_s | \to | N z_s$ only for the sake of completeness, since this is required by Definition 12.1. In the following we shall often omit instructions that cannot be executed.

Example 12.2 Let $\Sigma = \{a\}$ and $L = \emptyset$.

Again, we set $B = \{*, a, |\}$, $Z = \{z_s, z_f\}$. The desired Turing machine $M = [B, Z, A]$ is then completed by defining A as

$$z_s * \to *Nz_f \quad \text{and} \quad z_s a \to *Nz_f .$$

We finish with a more complicated example.

Example 12.3 Let $\Sigma = \{a, b\}$ and $L_{pal} = \{w \mid w \in \Sigma^*, w = w^T\}$.

For formally presenting a TM M accepting L_{pal} we set $B = \{*, |, a, b\}$, $Z = \{z_s, z_1, z_2, z_3, z_4, z_5, z_6, z_f\}$ and A is given in Figure 12.5.

	a	b	*	|
z_s	$*Rz_1$	$*Rz_2$	$|Nz_f$	$|Nz_f$
z_1	aRz_1	bRz_1	$*Lz_3$	$|Nz_f$
z_2	aRz_2	bRz_2	$*Lz_4$	$|Nz_f$
z_3	$*Lz_5$	$*Nz_f$	$|Nz_f$	$|Nz_f$
z_4	$*Nz_f$	$*Lz_5$	$|Nz_f$	$|Nz_f$
z_5	aLz_6	bLz_6	$|Nz_f$	$|Nz_f$
z_6	aLz_6	bLz_6	$*Rz_s$	$|Nz_f$

Figure 12.5 Instruction set of a TM accepting L_{pal}.

So, this machine remembers the actual leftmost and rightmost symbol, respectively. Then it is checking whether or not it is identical to the rightmost and leftmost symbol, respectively.

Exercise 12.1 *Construct a TM accepting L_{pal2}.*

Exercise 12.2 *Construct a TM accepting $L = \{a^n b^n c^n \mid n \in \mathbb{N}^+\}$.*

The following exercise is conceptually related to Corollary 12.1. If we fix any alphabet Σ, then it is also not hard to see that there is a universal TM that can simulate any other TM accepting a language over Σ. That is, given any coding of a TM M and a string w as input, the universal TM accepts w if and only if M accepts it. Now, we can define a universal DFA in the same way.

Exercise 12.3 *Prove or disprove: There is a universal DFA.*

12.5 Problem Set 12

Problem 12.1 Prove or disprove the following two assertions. In case your answer is affirmative, provide a Turing program.

(1) The binary addition function is Turing computable provided that the inputs are made in unary presentation.

(2) The binary multiplication function is Turing computable provided that the inputs are made in unary presentation.

Problem 12.2 Prove the following:

Let $n \in \mathbb{N}$, let $\tau \in \mathcal{T}^n$ and let $\psi \in \mathcal{T}^{n+2}$. Then we have: if

$$\phi(x_1, \ldots, x_n, 0) \quad = \tau(x_1, \ldots, x_n) \ ;$$
$$\phi(x_1, \ldots, x_n, y+1) = \psi(x_1, \ldots, x_n, y, \phi(x_1, \ldots, x_n, y)) \ ,$$

then $\phi \in \mathcal{T}^{n+1}$.

Problem 12.3 Prove the following:

Let $n \in \mathbb{N}^+$; then we have:

if $\tau \in \mathcal{T}^{n+1}$ and $\psi(x_1, \ldots, x_n) = \mu y[\tau(x_1, \ldots, x_n, y) = 1]$ then $\psi \in \mathcal{T}^n$.

13 Algorithmic Unsolvability

In the last chapter we have shown that there is a universal numbering $\psi \in \mathcal{P}^2$ for $\mathcal{P}$ (cf. Theorem 12.3). Furthermore, we have provided a TM U which is universal for $\mathcal{P}$ (= $\mathcal{T}$) (cf. Corollary 12.1). Now, we use these results to establish the first problem which is not algorithmically solvable.

13.1 The Halting Problem

Next, we explain the problem. Our universal TM U is computing every function in our universal numbering $\psi \in \mathcal{P}^2$. Now, if we wish to know whether or not $\psi(i, x)$ is defined, we could run U on input i, x. Then, two cases have to be distinguished. First, the computation of $U(i, x)$ stops. Then we know that $\psi(i, x)$ is defined, i.e., $\psi(i, x) \downarrow$. Second, the computation of $U(i, x)$ does *not* stop.

However, while everything is clear when the computation of $U(i, x)$ stops, the situation is different if it has not yet stopped. Just by observing the TM U on input i, x for a finite amount of time, the only thing we can tell for sure is that it has not *yet* stopped. Of course, there is still a chance that it will stop *later*. But it is also possible that it will *never* stop. Compare this to the situation when we are working with a computer. We have started a program and it does not terminate. What should we do? If we kill the execution of the program, then maybe it would have terminated its execution within the next hour, and now everything is lost. But if we let it run, and check the next day, and it still did not stop, again we have no idea what is better, to wait or to kill the execution.

Thus, it would be very nice if we could construct an algorithm *deciding* whether or not the computation of $U(i, x)$ will stop. So, we ask whether or not such an algorithm does exist. This problem is usually referred to as the *general halting problem* (一般停止問題). This equivalent to asking whether or not the following function $\tilde{h}\colon \mathbb{N} \times \mathbb{N} \to \mathbb{N}$ is computable, where

$$\tilde{h}(i, x) = \begin{cases} 1, & \text{if } \psi(i, x) \downarrow; \\ 0, & \text{if } \psi(i, x) \uparrow. \end{cases}$$

Clearly, $\tilde{h}$ is total. Thus, we have to figure out whether or not $\tilde{h} \in \mathcal{R}^2$.

For answering this question, we look at a restricted version of it, usually just called the *halting problem* (停止問題), i.e., we consider

$$h(x) = \begin{cases} 1, & \text{if } \psi(x, x) \downarrow; \\ 0, & \text{if } \psi(x, x) \uparrow. \end{cases}$$

Thus, now we have to find out whether or not $h \in \mathcal{R}$. The negative answer is provided by our next theorem. We shall even proof a stronger result which is independent of the particular choice of ψ.

Theorem 13.1 *Let $\psi \in \mathcal{P}^2$ be a universal numbering for $\mathcal{P}$. Then for*

$$h(x) = \begin{cases} 1, & \text{if } \psi(x, x) \downarrow; \\ 0, & \text{if } \psi(x, x) \uparrow, \end{cases}$$

we always have $h \notin \mathcal{R}$.

Proof. The proof is done by *diagonalization* (対角化). Suppose the converse, i.e., $h \in \mathcal{R}$. Now, we define a function $\overline{h}\colon \mathbb{N} \to \mathbb{N}$ as follows.

$$\overline{h}(x) = \begin{cases} 1 \dotdiv \psi(x, x), & \text{if } h(x) = 1; \\ 0, & \text{if } h(x) = 0. \end{cases}$$

13.1 The Halting Problem

Since $h \in \mathcal{R}$ by supposition, we can directly conclude that $\overline{h} \in \mathcal{R}$, too, by using the same technique as in the proof of Theorem 11.3. Furthermore, $\overline{h} \in \mathcal{R}$ directly implies $\overline{h} \in \mathcal{P}$, since $\mathcal{R} \subseteq \mathcal{P}$. Since ψ is universal for $\mathcal{P}$, there must exist an $i \in \mathbb{N}$ such that $\overline{h} = \psi_i$. Consequently, $\psi_i(i) = \overline{h}(i)$. We distinguish the following cases.

Case 1. $\psi_i(i) \downarrow$.

Then, by the definition of the function h we directly get that $h(i) = 1$. Therefore, the definition of the function $\overline{h}$ implies

$$\overline{h}(i) = 1 \dot{-} \psi(i,i) = 1 \dot{-} \psi_i(i) \neq \psi_i(i),$$

a contradiction to $\psi_i(i) = \overline{h}(i)$. Thus, Case 1 cannot happen.

Case 2. $\psi_i(i) \uparrow$.

Now, the definition of the function h directly implies that $h(i) = 0$. Therefore, the definition of the function $\overline{h}$ yields $\overline{h}(i) = 0$. So, we have

$$0 = \overline{h}(i) = \psi_i(i),$$

a contradiction to $\psi_i(i) \uparrow$. Thus, Case 2 cannot happen either.

Now, the only resolution is that our supposition $h \in \mathcal{R}$ must be wrong. Consequently, we get $h \notin \mathcal{R}$, and the theorem is proved. ∎

So, we have shown that the halting problem is *algorithmically unsolvable* (計算論的に解決不能), or as we say synonymously, *undecidable* (決定不能). Thus, we have seen that there is a concrete problem that cannot be solved by a computer, now or ever.

This directly allows for the corollary that the general halting problem is algorithmically unsolvable, too.

Corollary 13.1 *Let $\psi \in \mathcal{P}^2$ be a universal numbering for $\mathcal{P}$. Then for*

$$\tilde{h}(i,x) = \begin{cases} 1, & \text{if } \psi(i,x) \downarrow\,; \\ 0, & \text{if } \psi(i,x) \uparrow, \end{cases}$$

we always have $\tilde{h} \notin \mathcal{R}^2$.

Proof. Suppose the converse, i.e., $\tilde{h} \in \mathcal{R}^2$. Then, we directly see that $h(x) = \tilde{h}(x,x)$ and thus, $h \in \mathcal{R}$, too. But this is a contradiction to Theorem 13.1. The corollary follows. ∎

Next, we aim to attack some of the problems from formal language theory. But this is easier said than done. First, we have to study another problem which turns out to be very helpful, i.e., Post's [29] correspondence problem.

13.2 Post's Correspondence Problem

While we have provided a direct proof for the undecidability of the halting problem, in the following we use a different approach. That is, we *reduce* undecidable problems concerning Turing machines to "real" problems.

It is common to associate a problem with a set. Let $A \subseteq \mathbb{N}$ be any set. We define the *characteristic function* (特性関数) $\chi_A : A \to \{0,1\}$ as follows:

$$\chi_A(x) = \begin{cases} 1, & \text{if } x \in A\ ; \\ 0, & \text{if } x \notin A\ . \end{cases} \qquad (13.1)$$

Definition 13.1 *Let* $A \subseteq \mathbb{N}$; *then* A *is called* decidable (決定可能) *if* $\chi_A \in \mathcal{R}$. *If* A *is not decidable then we call it* undecidable (決定不能).

If a set A is decidable, then we also say that A is *recursive* (帰納的), otherwise, we also call it *not recursive* (非帰納的).

Next, let us explain what is meant by *reduction* (帰着). Let $A, B \subseteq \mathbb{N}$ and assume $\chi_A \in \mathcal{R}$. So we know A to be decidable. But our new problem is to decide for every $x \in \mathbb{N}$ whether or not $x \in B$. Instead of starting from scratch, we also have the possibility to search for a function *red* having the following two properties:

(1) $red \in \mathcal{R}$,

(2) $x \in B$ if and only if $red(x) \in A$.

Property (1) ensures that *red* is computable and total. Given any $x \in \mathbb{N}$, we compute $red(x)$ and then $\chi_A(red(x))$. By Property (2) we know that $x \in B$ if and only if $\chi_A(red(x)) = 1$ (cf. Figure 13.1). So, if we can find *red*, we know $\chi_B \in \mathcal{R}$. That is, we can decide B.

If B is reducible to A via *red* then we write $B \leq_{red} A$.

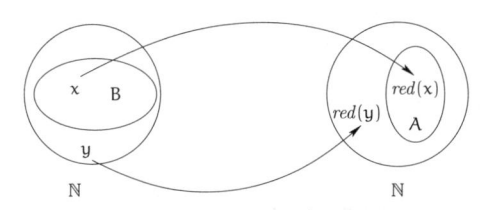

Figure 13.1 Reducing B to A.

Reductions are important as a proof technique, too. Let $\psi \in \mathcal{P}^2$ be any universal numbering for $\mathcal{P}$. The set corresponding to the halting problem is $K = \{i \mid i \in \mathbb{N}, \psi_i(i) \downarrow\}$. We refer to K as the *halting set* (停止集合). As shown in Theorem 13.1, we have $\chi_K \notin \mathcal{R}$. Thus, K is *undecidable*.

Now, let a set $P \subseteq \mathbb{N}$ be given. If we can find a reduction function $red \in \mathcal{R}$ such that $K \leq_{red} P$, then P must be undecidable, too. For seeing this, suppose the converse, i.e., $\chi_P \in \mathcal{R}$. Given any $x \in \mathbb{N}$, we can compute $\chi_P(red(x))$. By Property (2) of a reduction function, we then have $x \in K$ if and only if $\chi_P(red(x)) = 1$. Thus, K would be decidable, a contradiction.

We continue with Post's correspondence problem (ポストの対応問題). Though Post's correspondence problem may seem rather abstract to us, it has the advantage to be defined over strings. As we shall see, this allows for many applications of Post's correspondence problem to problems from formal language theory. We abbreviate Post's correspondence problem by PCP. Next, we formally define what a PCP is.

13. Algorithmic Unsolvability

Definition 13.2 *A quadruple* $[\Sigma, n, \mathfrak{P}, \mathfrak{Q}]$ *is called* PCP *if*

(1) $\Sigma \neq \emptyset$ *is a finite alphabet,*

(2) $n \in \mathbb{N}^+$,

(3) $\mathfrak{P}, \mathfrak{Q} \in (\Sigma^+)^n$, *i.e.,*

$$\mathfrak{P} = [p_1, \ldots, p_n]$$

$$\mathfrak{Q} = [q_1, \ldots, q_n], \quad \text{where } p_i, q_i \in \Sigma^+ .$$

Definition 13.3 *Let* $[\Sigma, n, \mathfrak{P}, \mathfrak{Q}]$ *be any PCP. The PCP* $[\Sigma, n, \mathfrak{P}, \mathfrak{Q}]$ *is said to be* solvable (解決可能) *if there is a finite sequence* $i_1, i_2 \ldots, i_k$ *of natural numbers such that*

(1) $i_j \leq n$ *for all* $1 \leq j \leq k$,

(2) $p_{i_1} p_{i_2} \cdots p_{i_k} = q_{i_1} q_{i_2} \cdots q_{i_k}$.

The PCP $[\Sigma, n, \mathfrak{P}, \mathfrak{Q}]$ *is* unsolvable (解決不能) *if it is not solvable.*

We give an example for a solvable and an unsolvable PCP, respectively.

Example 13.1 Consider the PCP $[\{a, b\}, 3, [a^2, b^2, ab^2], [a^2b, ba, b]]$. This PCP is solvable, since

$$p_1 p_2 p_1 p_3 = a^2 b^2 a^2 ab^2 = a^2 bbaa^2 bb = q_1 q_2 q_1 q_3 .$$

Example 13.2 Consider the PCP $[\{a\}, 2, [a^3, a^4], [a^2, a^3]]$.

This PCP is unsolvable (see also Problem 13.2).

Furthermore, it is necessary to have the following definition.

Definition 13.4 *Let* $[\Sigma, n, \mathfrak{P}, \mathfrak{Q}]$ *be any PCP. The PCP* $[\Sigma, n, \mathfrak{P}, \mathfrak{Q}]$ *is said to be* 1-solvable (*1-解決可能*) *if it is solvable and*

$$p_1 p_{i_2} \cdots p_{i_k} = q_1 q_{i_2} \cdots q_{i_k} ,$$

that is, if $i_1 = 1$.

Now we are ready to state the main theorem of this section, i.e., the property of a PCP to be solvable is undecidable.

13.2 Post's Correspondence Problem

Theorem 13.2 *There does not exist any Turing machine* M *such that*

$$M(p_1 \# \cdots \# p_n \# q_1 \# \cdots \# q_n) = \begin{cases} 1, & \text{if } [\Sigma, n, \mathfrak{P}, \mathfrak{Q}] \text{ is solvable;} \\ 0, & \text{if } [\Sigma, n, \mathfrak{P}, \mathfrak{Q}] \text{ is unsolvable,} \end{cases}$$

for every PCP $[\Sigma, n, \mathfrak{P}, \mathfrak{Q}]$.

Proof. The proof idea is as follows. We shall show that deciding the solvability of any PCP is at least as hard as deciding the halting problem. This goal is achieved in two steps, i.e., we show:

Step 1. If we could decide the solvability of any PCP, then we can also decide the 1-solvability of any PCP.

Step 2. If we could decide the 1-solvability of any PCP, then we can also decide the halting problem.

For proving Step 1, we show that for every PCP $[\Sigma, n, \mathfrak{P}, \mathfrak{Q}]$ we can effectively construct a PCP $[\Sigma \cup \{A, E\}, n + 2, \mathfrak{P}', \mathfrak{Q}']$ such that

$[\Sigma, n, \mathfrak{P}, \mathfrak{Q}]$ is 1-solvable iff

$$[\Sigma \cup \{A, E\}, n + 2, \mathfrak{P}', \mathfrak{Q}'] \text{ is solvable .} \tag{13.2}$$

Here, by *effective* we mean that there is an algorithm (a Turing machine) which on input any PCP $[\Sigma, n, \mathfrak{P}, \mathfrak{Q}]$ outputs $[\Sigma \cup \{A, E\}, n + 2, \mathfrak{P}', \mathfrak{Q}']$.

We define the following two homomorphisms h_R and h_L. Let A, E be any fixed symbols such that $A, E \notin \Sigma$. Then we set for all $x \in \Sigma$

$$h_R(x) = xA, \quad \text{and} \quad h_L(x) = Ax . \tag{13.3}$$

Next, let PCP $[\Sigma, n, \mathfrak{P}, \mathfrak{Q}]$ be given as input, where $\mathfrak{P} = [p_1, \ldots, p_n]$ and $\mathfrak{Q} = [q_1, \ldots, q_n]$. Then we compute $\mathfrak{P}'$ and $\mathfrak{Q}'$ as follows:

$$\begin{aligned} p_1' &= Ah_R(p_1) & q_1' &= h_L(q_1) \\ p_{i+1}' &= h_R(p_i) \quad 1 \le i \le n & q_{i+1}' &= h_L(q_i) \\ p_{n+2}' &= E & q_{n+2}' &= AE \end{aligned}$$

13. Algorithmic Unsolvability

For the sake of illustration, let us exemplify the construction. Let the PCP $[\{a, b\}, 2, [a^2, b], [a, ab]]$ be given. It is obviously 1-solvable, since $p_1 p_2 = q_1 q_2$. Next, we compute $\mathfrak{P}'$ and $\mathfrak{Q}'$ as follows:

$$p'_1 = AaAaA \qquad q'_1 = Aa$$
$$p'_2 = aAaA \qquad q'_2 = Aa$$
$$p'_3 = bA \qquad q'_3 = AaAb$$
$$p'_4 = E \qquad q'_4 = AE$$

Now, it is easy to see that $[\{a, b, A, E\}, 4, \mathfrak{P}', \mathfrak{Q}']$ is solvable, since

$$p'_1 p'_3 p'_4 = AaAaAbAE = q'_1 q'_3 q'_4 .$$

For the general case, we show the following claim stating that if $[\Sigma, n, \mathfrak{P}, \mathfrak{Q}]$ is 1-solvable then $[\Sigma \cup \{A, E\}, n + 2, \mathfrak{P}', \mathfrak{Q}']$ is solvable.

Claim 1. *Let $[\Sigma, n, \mathfrak{P}, \mathfrak{Q}]$ be any PCP and let $[\Sigma \cup \{A, E\}, n + 2, \mathfrak{P}', \mathfrak{Q}']$ be the PCP constructed as described above. If there is a finite sequence $1, i_1, \ldots, i_r$ such that $p_1 p_{i_1} \cdots p_{i_r} = q_1 q_{i_1} \cdots q_{i_r}$ then*
$$p'_1 p'_{i_1+1} \cdots p'_{i_r+1} p'_{n+2} = q'_1 q'_{i_1+1} \cdots q'_{i_r+1} q'_{n+2}.$$

The claim is proved by calculating the strings involved. For seeing how the proof works, let us start with the simplest case, i.e., $p_1 = q_1$.

Let $p_1 = x_1 \cdots x_k$. By construction we get

$$p'_1 = Ax_1 Ax_2 A \cdots Ax_k A \qquad q'_1 = Ax_1 Ax_2 A \cdots Ax_k$$
$$p'_2 = x_1 Ax_2 A \cdots Ax_k A \qquad q'_2 = Ax_1 Ax_2 A \cdots Ax_k$$
$$p'_{n+2} = E \qquad q'_{n+2} = AE$$

So, we see that p'_2 and q'_2 are almost equal, that is they are equal except the missing leading A in p'_2 and the missing A at the end of q'_2. Therefore, we replace p'_2 and q'_2 by p'_1 and q'_1, respectively. This solves the problem of the missing leading A. Now, q'_{n+2} gives the missing A on the rightmost part of q'_1 plus an E and this E is obtained from p'_{n+2} which is concatenated to p'_2. Hence, $p'_1 p'_{n+2} = q'_1 q'_{n+2}$. Thus the assertion follows.

13.2 Post's Correspondence Problem

Now, for the general case the same idea works. Assume that

$$p_1 p_{i_1} \cdots p_{i_r} = q_1 q_{i_1} \cdots q_{i_r}. \tag{13.4}$$

We have to show that

$$p_1' p_{i_1+1}' \cdots p_{i_r+1}' p_{n+2}' = q_1' q_{i_1+1}' \cdots q_{i_r+1}' q_{n+2}'.$$

Since h_R and h_L are homomorphisms, by construction we get

$$h_R(p_1 p_{i_1} \cdots p_{i_r}) = h_R(p_1) h_R(p_{i_1}) \cdots h_R(p_{i_r})$$
$$= p_2' p_{i_1+1}' \cdots p_{i_r+1}', \quad \text{and}$$
$$h_L(q_1 q_{i_1} \cdots q_{i_r}) = h_L(q_1) h_L(q_{i_1}) \cdots h_L(q_{i_r})$$
$$= q_2' q_{i_1+1}' \cdots q_{i_r+1}'.$$

Thus by (13.4), applying h_R to $p_1 p_{i_1} \cdots p_{i_r}$ and h_L to $q_1 q_{i_1} \cdots q_{i_r}$ gives again almost the same strings, except the same problems mentioned above. So we again replace p_2' and q_2' by p_1' and q_1', respectively, solving the problem of the leading A in $h_R(p_1 p_{i_1} \cdots p_{i_r})$. Appending p_{n+2}' and q_{n+2}' to $p_1' p_{i_1+1}' \cdots p_{i_r+1}'$ and $q_1' q_{i_1+1}' \cdots q_{i_r+1}'$, respectively, gives on the left hand side an E at the end of the string and an AE on the right hand side at the end of the string. Hence,

$$p_1' p_{i_1+1}' \cdots p_{i_r+1}' p_{n+2}' = q_1' q_{i_1+1}' \cdots q_{i_r+1}' q_{n+2}'. \tag{13.5}$$

This proves the claim.

Thus, we have shown the necessity of (13.2). For proving the sufficiency of (13.2), assume that $[\Sigma \cup \{A, E\}, n+2, \mathfrak{P}', \mathfrak{Q}']$ is solvable. We have to show that $[\Sigma, n, \mathfrak{P}, \mathfrak{Q}]$ is 1-solvable.

By assumption, there is a finite sequence $i_1, \ldots, i_r$ such that

$$p_{i_1}' \cdots p_{i_r}' = q_{i_1}' \cdots q_{i_r}'. \tag{13.6}$$

Now, all strings q'_i start with an A. Since p'_1 is the only string among all the strings p'_i which starts with A, we directly see that $i_1 = 1$.

Furthermore, since all p'_i end with an A and none of the q'_i does end with A, we can directly conclude that $i_r = n + 2$. Thus, by deleting all A and E we directly get the desired solution, i.e.,

$$p_1 p_{i_2-1} \cdots p_{i_r-1} = q_1 q_{i_2-1} \cdots q_{i_r-1} . \tag{13.7}$$

Note that deleting all A and E corresponds to applying the erasing homomorphism h_e defined as $h_e(x) = x$ for all $x \in \Sigma$ and $h_e(A) = h_e(E) = \lambda$. Therefore, we have provided a 1-solution of $[\Sigma, n, \mathfrak{P}, \mathfrak{Q}]$. This completes the proof of Step 1.

We continue with Step 2. The idea is to transform the instruction set of any TM M and its actual input x into a PCP such that the PCP is 1-solvable if and only if M stops its computation on input x.

For realizing this idea, it is technically advantageous to assume a normalization of TMs such as we did when proving Lemma M^+ in Chapter 12. Therefore, we need the following Lemma M^*.

Lemma M^*. *For every TM M there exists a TM M^* such that*

(1) $f_M = f_{M^*}$, *i.e., M and M^* compute the same function.*

(2) M^* *is never moving left to the initial cell observed when starting its computation.*

(3) *In each step of the computation of M^* the head is either moving one cell to the left or one cell to the right.*

(4) M^* *is not allowed to write the symbol $*$ on its tape (but of course, it is allowed to read it).*

Proof. Property (2) of Lemma M^* can be shown in the same way as we proved it in the demonstration of Lemma M^+. But instead of L, now we use another symbol, e.g., **L**.

13.2 Post's Correspondence Problem

For showing Property (3), we have to deal with all instructions of the form $zb \to b'Nz'$ of M's instruction set. So, if the instruction $zb \to b'Nz'$ belongs to M's instruction set, then we introduce a new state $z_{z'}$ to the state set of M^* and replace $zb \to b'Nz'$ by

$zb \to b'Lz_{z'}$

$z_{z'}b \to bRz'$ for all tape symbols $b \in B$.

That is, now the head first moves left and M^* changes its state to $z_{z'}$. And then it moves the head just right and switches it state back to z'. Thus, it behaves exactly as M does except the two additional moves of the head.

Finally, we have to realize Property (4). Without loss of generality we can assume that $? \notin B$. Then instead of writing $*$ on its tape, M^* writes the new symbol ? on its tape. That is, we replace all instructions $zb \to *Hz'$, where $H \in \{L, R\}$, of the so far transformed instruction set by $zb \to ?Hz'$, where $H \in \{L, R\}$. Additionally, we have to duplicate all instructions of the form $z* \to \ldots$ by $z? \to \ldots$. That is, when reading a ?, M^* works as M would do when reading a $*$.

Now, it is easy to see that Property (1) is also satisfied. We omit details. This proves Lemma M^*.

Introducing the ? in Lemma M^* has the advantage that M^*, when reading a $*$ knows that it is visiting a cell it has not visited before.

In the same way as we did in Chapter 12, one can show that there is a universal TM U^* which can simulate all TMs M^*. Since the TMs M^* have special properties, it is conceivable that the halting problem for U^* is decidable. However, using the same technique as above, one can show the general halting problem and the halting problem for U^* to be undecidable.

This is a good point to recall that we have introduced instantaneous descriptions for pushdown automata in order to describe their computations.

13. Algorithmic Unsolvability

For describing computations of Turing machines we shall use *configurations* (様相) defined as follows. Let M^* be a TM.

A string $b_1 b_2 \cdots b_{i-1} z b_i \cdots b_n$ is said to be a *configuration* of M^* if

1. z is the actual state of M^*,
2. the head is observing cell i,
3. $b_1 b_2 \cdots b_n$ is the portion of the tape between the leftmost and rightmost $*$.

So, the initial configuration is $z_s x$, where x represents the input. If we use an unary representation for the inputs, then, in case $x = 0$, we omit the $*$. We write $c_i \vdash c_{i+1}$ provided configuration c_{i+1} is reached from configuration c_i in one step of computation performed by M^*.

Example 13.3 Let $M_{ex}^* = [\{*, |\}, \{z_s, z_f, z_1\}, A]$, where A is defined as:

$$z_s | \to |R z_1 \qquad\qquad z_s * \to |R z_f$$
$$z_1 | \to |R z_1 \qquad\qquad z_1 * \to |L z_f$$

Then, on input $|$ we get the following sequence of configurations.

$$z_s | \vdash | z_1 \vdash z_f | | \tag{13.8}$$

Now, we perform the final step of our proof, i.e., reducing the halting problem to PCP. Given any TM $M^* = [B, Z, A]$ and input x represented unary, we construct a PCP as shown in Figure 13.2. By Lemma M^* we have that if $zb \to z'Hb' \in A$, where $H \in \{L, R\}$, then $b' \in B \setminus \{*\}$. In Figure 13.2, we always assume that b, b', $b'' \in B \setminus \{*\}$. Also, we assume that $\# \notin B$, i.e., here we use $\#$ as a separator (of configurations).

Before continuing with the proof, we return to Example 13.3. Note that $f_{M_{ex}^*}(0) = 0$ and $f_{M_{ex}^*}(n) = 2$ for all $n \in \mathbb{N}^+$. So, M_{ex}^* stops on every input.

Then for any $x \in \mathbb{N}$, we get the following PCP, where again x is the input to M_{ex}^* represented in unary notation (cf. Figure 13.3).

13.2 Post's Correspondence Problem

	$\mathfrak{P}$		$\mathfrak{Q}$	if
p_1	#	q_1	$\#z_s\| \cdots \|\#$	$x = \| \cdots \|$
	zb		$b'z$	$zb \to b'Rz' \in A$
	bzb'		$z'bb''$	$zb' \to b''Lz' \in A, b \in B \setminus \{*\}$
	$z\#$		$bz'\#$	$z* \to bRz' \in A$
p_i	$bz\#$	q_i	$zbb'\#$	$z* \to b'Lz' \in A, b \in B \setminus \{*\}$
	b		b	$b \in B \setminus \{*\}$
	#		#	always
	$z_f b$		z_f	$b \in B \setminus \{*\}$
	bz_f		z_f	$b \in B \setminus \{*\}$
	$z_f\#\#$		#	always

Figure 13.2 The PCP corresponding to M^* on input x.

	$\mathfrak{P}$		$\mathfrak{Q}$	if
p_1	#	q_1	$\#z_s\|\cdots\|\#$	$x = \|\cdots\|$
p_2	$z_s\|$	q_2	$\|z_1$	$z_s\| \to \|Rz_1 \in A$
p_3	$z_1\|$	q_3	$\|z_1$	$z_1\| \to \|Rz_1 \in A$
p_4	$z_s\#$	q_4	$\|z_f\#$	$z_s * \|Rz_f \in A$
p_5	$\|z\#$	q_5	$z_f\|\|\#$	$z_1* \to \|Lz_f \in A$
p_6	$\|$	q_6	$\|$	$\| \in B \setminus \{*\}$
p_7	#	q_7	#	always
p_8	$z_f\|$	q_8	z_f	$\| \in B \setminus \{*\}$
p_9	$\|z_f$	q_9	z_f	$\| \in B \setminus \{*\}$
p_{10}	$z_f\#\#$	q_{10}	#	always

Figure 13.3 The PCP corresponding to M^*_{ex} on input x.

Looking at Figure 13.3, we see that $|p_i| \leq |q_i|$ for all $i \in \{1, \ldots, 7\}$. In order to find a 1-solution of the PCP we concatenate p's and q's, respectively, that correspond to the computation of M^*_{ex} when started in z_s on input x. Thus, initially, the concatenation of the p's will result in a string that is shorter than the string obtained when concatenating the corresponding q's. This length difference can be resolved if and only if the computation terminates. Then, one can use the strings from the last group in Figure 13.3 to resolve the length difference.

For seeing how it works, we consider input $|$. Therefore, $p_1 = \#$ and $q_1 = \#z_s|\#$. Hence, in order to have any chance to find a 1-solution of the PCP, we must use p_2 next. Consequently, we have to use q_2, too, and get

$$\underbrace{\#}_{p_1} \underbrace{z_s\,|}_{p_2}$$
$$\underbrace{\#z_s\,|\#}_{q_1} \underbrace{|z_1}_{q_2}$$

Next, in the sequence of the p's, we need a $\#$. We could use p_1 or p_7. So, we try p_7, and hence, we have to use q_7, too. Then we must use p_5 and q_5. Thus, we have

$$\underbrace{\#}_{p_1} \underbrace{z_s\,|}_{p_2} \underbrace{\#}_{p_7} \underbrace{|z_1\#}_{p_5}$$
$$\underbrace{\#z_s\,|\#}_{q_1} \underbrace{|z_1}_{q_2} \underbrace{\#}_{q_7} \underbrace{z_f\,||\#}_{q_5}$$

Now, the idea of our construction should be clear. Before reaching the final state, the sequence of q's describes successive configurations of the computation of M_{ex}^* started on input x. That is, if we have $\#\alpha_i\#\alpha_{i+1}$ then α_i and α_{i+1} are configurations of M_{ex}^* and $\alpha_i \vdash \alpha_{i+1}$ holds (cf. (13.8)). Also note that the sequence of q's is always one configuration ahead to the sequence generated by the p's as long as M_{ex}^* did not reach its final state.

Looking at the sequences generated so far, we see that after z_f has been reached the strings p_8, p_9 and p_{10} can be used, too. So, we get the following 1-solution by appending $p_8p_6p_7p_8p_7p_{10}$ to the already used sequence $p_1p_2p_7p_5$ and $q_8q_6q_7q_8q_7q_{10}$ to $q_1q_2q_7q_5$ resulting in

$$p_1p_2p_7p_5p_8p_6p_7p_8p_7p_{10} = \#z_s\,|\#|z_1\#z_f\,||\#z_f\,|\#z_f\#\#$$

$$q_1q_2q_7q_5q_8q_6q_7q_8q_7q_{10} = \#z_s\,|\#|z_1\#z_f\,||\#z_f\,|\#z_f\#\#$$

Next, we finish the proof by showing the following claim.

Claim 2. The computation of M^* *on input* x *stops if and only if the corresponding PCP is 1-solvable.*

We start with the sufficiency. If the corresponding PCP is 1-solvable then the solution must start with p_1 and q_1. By construction (cf. Figure 13.2), we have $p_1 = \#$ and $q_1 = \#z_s|\cdots|\#$. Looking at Figure 13.2, we see that $|p_i| \leq |q_i|$ for all strings except the ones in the last group. Thus, as long as M^* does not reach its final state, the concatenation of p's is shorter than the concatenation of q's. Furthermore, $z_s|\cdots|$ corresponds to the initial configuration of M^* on input x. By construction, if the sequence of p's is s and the sequence of q's is sy then s is a sequence of configurations of M^* representing the computation of M^* on input x possibly followed by # and the beginning of the next configuration. Consequently, if the PCP is 1-solvable, then M^* has reached its final state. This proves the sufficiency.

Necessity. If M^* on input x reaches its final state, then, after having started with p_1 and q_1, we can use the p_i's and the corresponding q_i's from the first and second group to obtain the sequence of configurations as well as the separator #. Having done this, we finally use the p_i's and the corresponding q_i's from the second and third group to get the desired solution of the PCP. ∎

Note that Theorem 13.2 does *not* hold if $\text{card}(\Sigma) = 1$ (see Problem 13.2).

13.3 Problem Set 13

Problem 13.1 Let $f, g: \mathbb{N} \to \mathbb{N}$ be functions. Function f is a subfunction of g (written $f \subseteq g$) if for all $x \in \mathbb{N}$ we have: If $f(x) \downarrow$ then $f(x) = g(x)$. Prove the following:

There is a function $f \in \mathcal{P}$ such that there is no function $g \in \mathcal{R}$ with $f \subseteq g$.

Problem 13.2 Prove the following:

PCP $[\Sigma, n, \mathfrak{P}, \mathfrak{Q}]$ is decidable for all n provided $\text{card}(\Sigma) = 1$.

Problem 13.3 Provide a 1-solution for the PCP obtained for M^*_{ex} when started on input || (cf. Figure 13.3).

14 Applications of PCP

Our goal is to present some typical undecidability results for problems arising naturally in formal language theory. The main focus is on context-free languages but we also look at regular languages and the family $\mathcal{L}_0$.

14.1 Results for Context-Free Languages

As we have seen in Chapter 6, there are context-free languages L_1 and L_2 such that $L_1 \cap L_2 \notin \mathcal{CF}$. So, it would be nice to have an algorithm which, on input any two context-free grammars $\mathcal{G}_1$, $\mathcal{G}_2$, returns 1 if $L(\mathcal{G}_1) \cap L(\mathcal{G}_2) \in \mathcal{CF}$ and 0 otherwise. Unfortunately, such an algorithm does not exist. Also, some closely related problems cannot be solved algorithmically.

Theorem 14.1 *The following problems are undecidable for any context-free grammars $\mathcal{G}_1$, $\mathcal{G}_2$:*

(1) $L(\mathcal{G}_1) \cap L(\mathcal{G}_2) = \emptyset$,

(2) $L(\mathcal{G}_1) \cap L(\mathcal{G}_2)$ *is infinite,*

(3) $L(\mathcal{G}_1) \cap L(\mathcal{G}_2) \in \mathcal{CF}$,

(4) $L(\mathcal{G}_1) \cap L(\mathcal{G}_2) \in \mathcal{REG}$.

Proof. The general proof idea is as follows. We construct a context-free language L_S (here S stands for standard) and for any PCP $[\{a, b\}, n, \mathfrak{P}, \mathfrak{Q}]$ a language $L(\mathfrak{P}, \mathfrak{Q})$ such that $L_S \cap L(\mathfrak{P}, \mathfrak{Q}) \neq \emptyset$ iff $[\{a, b\}, n, \mathfrak{P}, \mathfrak{Q}]$ is solvable.

14.1 Results for Context-Free Languages

Let $\Sigma = \{a, b, c\}$ and define

$$L_S = \{pcqcq^Tcp^T \mid p, q \in \{a, b\}^+, c \in \Sigma\} . \tag{14.1}$$

Furthermore, for any $p_1, \ldots, p_n$, where $p_i \in \{a, b\}^+$, we set

$$L(p_1, \ldots, p_n) =$$
$$\{ba^{i_k}b \cdots ba^{i_1}cp_{i_1}p_{i_2} \cdots p_{i_k} \mid k \geq 1, \forall j[1 \leq j \leq k \implies 1 \leq i_j \leq n]\} .$$

Here the idea is to encode in a^{i_j} the index i_j. Now, let $[\{a, b\}, n, \mathfrak{P}, \mathfrak{Q}]$ be any PCP, then we define the language $L(\mathfrak{P}, \mathfrak{Q})$ as follows:

$$L(\mathfrak{P}, \mathfrak{Q}) = L(p_1, \ldots, p_n)\{c\}L^T(q_1, \ldots, q_n) . \tag{14.2}$$

Claim 1. L_S, $L(p_1, \ldots, p_n)$ and $L(\mathfrak{P}, \mathfrak{Q})$ *are context-free.*

First, we define a grammar $\mathcal{G}_S = [\{a, b, c\}, \{\sigma, h\}, \sigma, P]$, where the production set P is defined as follows:

$\sigma \to a\sigma a$	$\sigma \to chc$	$h \to bhb$
$\sigma \to b\sigma b$	$h \to aha$	$h \to c$

Now, it is easy to see that $\mathcal{G}_S$ is context-free and that

$$\sigma \overset{*}{\Rightarrow} w\sigma w^T \Rightarrow wchcw^T \overset{*}{\Rightarrow} wcvhv^Tcw^T \Rightarrow wcvcv^Tcw^T ,$$

where $w, v \in \{a, b\}^+$. Hence $L(\mathcal{G}_S) \subseteq L_S$. The inclusion $L_S \subseteq L(\mathcal{G}_S)$ is obvious. Consequently, $L_S \in \mathcal{CF}$.

Let $\mathfrak{P} = [p_1, \ldots, p_n]$. We define a grammar $\mathcal{G}_\mathfrak{P} = [\{a, b, c\}, \{\sigma\}, \sigma, P]$, where $P = \{\sigma \to ba^i\sigma p_i \mid i = 1, \ldots, n\} \cup \{\sigma \to c\}$. Clearly, $\mathcal{G}_\mathfrak{P}$ is context-free, $L(\mathcal{G}_\mathfrak{P}) = L(p_1, \ldots, p_n)$ and thus $L(p_1, \ldots, p_n) \in \mathcal{CF}$.

By Theorem 6.3 we know that $L^T(p_1, \ldots, p_n) \in \mathcal{CF}$, too. Moreover, $\mathcal{CF}$ is closed under product (cf. Theorem 6.2). Consequently, $L(\mathfrak{P}, \mathfrak{Q})$ is context-free for any PCP. This proves Claim 1.

14. Applications of PCP

Claim 2. For every PCP we have: $L_S \cap L(\mathfrak{P}, \mathfrak{Q}) \neq \emptyset$ if and only if $[\{a, b\}, n, \mathfrak{P}, \mathfrak{Q}]$ is solvable.

Necessity. Let $L_S \cap L(\mathfrak{P}, \mathfrak{Q}) \neq \emptyset$. Consider any string $r \in L_S \cap L(\mathfrak{P}, \mathfrak{Q})$:

$$r = \underbrace{ba^{i_k}b \cdots ba^{i_1}}_{w_1} c \underbrace{p_{i_1} p_{i_2} \cdots p_{i_k}}_{w_3} c \underbrace{(q_{j_1} \cdots q_{j_m})^T}_{w_4} c \underbrace{(ba^{j_m}b \cdots ba^{j_1})^T}_{w_2}.$$

Since $r \in L_S$, we directly see that $w_1 = w_2^T$ and $w_3 = w_4^T$. Consequently, we get $k = m$ and $i_\ell = j_\ell$ for $\ell = 1, \ldots, k$. Thus, the equality $w_3 = w_4^T$ provides a solution of $[\{a, b\}, n, \mathfrak{P}, \mathfrak{Q}]$.

Sufficiency. Let $[\{a, b\}, n, \mathfrak{P}, \mathfrak{Q}]$ be solvable. Then there is a finite sequence $i_1, i_2 \ldots, i_k$ of natural numbers such that $i_j \leq n$ for all $1 \leq j \leq k$, and $p_{i_1} p_{i_2} \cdots p_{i_k} = q_{i_1} q_{i_2} \cdots q_{i_k}$. Therefore, one directly gets a string $r \in L_S \cap L(\mathfrak{P}, \mathfrak{Q})$. This proves Claim 2.

Claim 1 and 2 together directly imply Assertion (1) via Theorem 13.2.

To show Assertion (2), we use the same ideas plus the following Claim.

Claim 3. $L_S \cap L(\mathfrak{P}, \mathfrak{Q})$ is infinite if and only if $L_S \cap L(\mathfrak{P}, \mathfrak{Q}) \neq \emptyset$.

The necessity is trivial. For showing the sufficiency, let $i_1, i_2 \ldots, i_k$ be a finite sequence of natural numbers such that $i_j \leq n$ for all $1 \leq j \leq k$, and $p_{i_1} p_{i_2} \cdots p_{i_k} = q_{i_1} q_{i_2} \cdots q_{i_k}$. Therefore, we also have $(p_{i_1} p_{i_2} \cdots p_{i_k})^m = (q_{i_1} q_{i_2} \cdots q_{i_k})^m$, that is, $(i_1, i_2 \ldots, i_k)^m$ is a solution of $[\{a, b\}, n, \mathfrak{P}, \mathfrak{Q}]$ for every $m \geq 1$. But this means, if

$$w_1 c w_3 c w_4 c w_2 \in L_S \cap L(\mathfrak{P}, \mathfrak{Q}), \quad \text{then also}$$

$$w_1^m c w_3^m c w_4^m c w_2^m \in L_S \cap L(\mathfrak{P}, \mathfrak{Q}) \quad \text{for every } m \in \mathbb{N}^+.$$

This proves Claim 3, and thus Assertion (2) is shown.

Next, we prove Assertions (3) and (4). This done via the following claim.

Claim 4. $L_S \cap L(\mathfrak{P}, \mathfrak{Q})$ does not contain any infinite context-free language.

Assuming Claim 4, Assertions (3) and (4) can be obtained, since the

14.1 Results for Context-Free Languages

following assertions are equivalent.

(α) $L_S \cap L(\mathfrak{P}, \mathfrak{Q}) = \emptyset$,

(β) $L_S \cap L(\mathfrak{P}, \mathfrak{Q}) \in \mathcal{REG}$,

(γ) $L_S \cap L(\mathfrak{P}, \mathfrak{Q}) \in \mathcal{CF}$.

Obviously, (α) implies (β) and (β) implies (γ). Thus, we have only to prove that (γ) implies (α). This is equivalent to showing that the negation of (α) implies the negation of (γ).

So, let us assume the negation of (α), i.e., $L_S \cap L(\mathfrak{P}, \mathfrak{Q}) \neq \emptyset$. By Claim 3, we then know that $L_S \cap L(\mathfrak{P}, \mathfrak{Q})$ is infinite. Now Claim 4 tells us that $L_S \cap L(\mathfrak{P}, \mathfrak{Q}) \notin \mathcal{CF}$. Thus, we have shown the negation of (γ).

Under the assumption that Claim 4 is true, we have thus established the equivalence of (α), (β) and (γ). Consequently, by Claim 2 we then know that $L_S \cap L(\mathfrak{P}, \mathfrak{Q}) \in \mathcal{CF}$ if and only if $[\{a, b\}, n, \mathfrak{P}, \mathfrak{Q}]$ is not solvable and also that $L_S \cap L(\mathfrak{P}, \mathfrak{Q}) \in \mathcal{REG}$ if and only if $[\{a, b\}, n, \mathfrak{P}, \mathfrak{Q}]$ is not solvable. This proves Assertions (3) and (4).

So, it remains to show Claim 4. Let $[\{a, b\}, n, \mathfrak{P}, \mathfrak{Q}]$ be arbitrarily fixed. Suppose there is a language $L \subseteq L_S \cap L(\mathfrak{P}, \mathfrak{Q})$ such that L is infinite and context-free. Now we apply Theorem 7.3. Thus, there exists a $k \in \mathbb{N}$ such that for all $w \in L$ with $|w| \geq k$ there are strings q, r, s, u, v such that $w = qrsuv$ and $ru \neq \lambda$ and $qr^i su^i v \in L$ for all $i \in \mathbb{N}$.

Since by supposition L is infinite, there must exist a string $w \in L$ with $|w| \geq k$. Furthermore, $L \subseteq L(\mathfrak{P}, \mathfrak{Q})$ and therefore w must have the form

$$w = w_1 c w_2 c w_3 c w_4 \,, \quad \text{where } w_i \in \{a, b\}^+, \ i = 1, 2, 3, 4\,. \quad (14.3)$$

That is, w contains exactly three times the letter c. Now, let $w = qrsuv$. We distinguish the following cases.

Case 1. c is a substring of ru.

Then $qr^i su^i v \in L$ for every $i \in \mathbb{N}$ and hence $qr^i su^i v$ contains at least four c's for every $i \geq 4$. So, Case 1 cannot happen.

Case 2. c is not a substring of ru.

Then, neither r nor u could be substrings of $ba^{i_1}b \cdots ba^{i_k}$. If r or u start and end with b, then $qr^2su^2v \notin L(\mathfrak{P}, \mathfrak{Q})$. If both r and u do not start and end with b, then $r \in \{a\}^+$ or $u \in \{a\}^+$ is impossible, since $qr^{n+1}su^{n+1}v$ then violates the condition that we have at most n consecutive a's. Otherwise, we get a contradiction to the definition of $L(\mathfrak{P}, \mathfrak{Q})$, since then we have more blocks of a's than there are p_i's or q_j's.

So, the only remaining possibility is that r is a substring of w_2 or r is a substring of w_3 (cf. (14.3)). In either case, then $qr^isu^iv \notin L(\mathfrak{P}, \mathfrak{Q})$ for $i \geq 2$, since the length of w_1 and w_2 as well as the length of w_3 and w_4 are related. This is one of the reasons we have included the a^{i_j} into the definition of $L(\mathfrak{P}, \mathfrak{Q})$. This proves Claim 4, and the theorem is shown. ∎

Next, we turn our attention to problems involving the complement of context-free languages. Recall that $\mathcal{CF}$ is not closed under complement (cf. Corollary 6.1). However, due to lack of space, we have to omit a certain part of the following proof.

Theorem 14.2 *The following problems are undecidable for any context-free grammar* $\mathcal{G}$:

(1) $\overline{L(\mathcal{G})} = \emptyset$,

(2) $\overline{L(\mathcal{G})}$ *is infinite*,

(3) $\overline{L(\mathcal{G})} \in \mathcal{CF}$,

(4) $\overline{L(\mathcal{G})} \in \mathcal{REG}$,

(5) $L(\mathcal{G}) \in \mathcal{REG}$.

Proof. We use the notions from the demonstration of Theorem 14.1. Consider $L =_{df} \overline{L_S \cap L(\mathfrak{P}, \mathfrak{Q})}$. Note that L is always context-free. We do not prove this assertion here. The reader is referred to Ginsburg [9].

14.1 Results for Context-Free Languages

For showing (1), suppose the converse. Then, given the fact that L is context-free, there is a context-free grammar $\mathcal{G}$ such that $L = L(\mathcal{G})$. So, we could run the algorithm on input $\mathcal{G}$. On the other hand, $\overline{L} = L_S \cap L(\mathfrak{P}, \mathfrak{Q})$. Thus, we could decide whether or not $L_S \cap L(\mathfrak{P}, \mathfrak{Q}) = \emptyset$. By Claim 2 in the proof of Theorem 14.1, this implies that we can decide whether or not PCP is solvable; a contradiction to Theorem 13.2.

Assertion (2) is shown analogously via (2) of Theorem 14.1.

Assertion (3) and (4) also follow directly from Theorem 14.1 by using its Assertions (3) and (4), respectively.

Finally, Assertion (5) is a direct consequence of Assertion (4) and the fact that $L \in \mathcal{REG}$ if and only if $\overline{L} \in \mathcal{REG}$ (cf. Problem 4.2). ∎

The theorems shown above directly allow for the following corollary.

Corollary 14.1 *The following problems are undecidable for any context-free grammars $\mathcal{G}_1$, $\mathcal{G}_2$:*

(1) $L(\mathcal{G}_1) = L(\mathcal{G}_2)$,

(2) $L(\mathcal{G}_1) \subseteq L(\mathcal{G}_2)$.

Proof. Suppose (1) is decidable. Let $\mathcal{G}_1$ be any context-free grammar such that $L(\mathcal{G}_1) = L = \overline{L_S \cap L(\mathfrak{P}, \mathfrak{Q})}$ (see the proof of Theorem 14.2). Let $\mathcal{G}_2$ be any context-free grammar such that $L(\mathcal{G}_2) = \{a, b, c\}^*$. Then

$$L = \{a,b,c\}^* \iff \overline{L_S \cap L(\mathfrak{P}, \mathfrak{Q})} = \{a,b,c\}^* \iff L_S \cap L(\mathfrak{P}, \mathfrak{Q}) = \emptyset$$

$$\iff [\{a,b\}, n, \mathfrak{P}, \mathfrak{Q}] \text{ is not solvable },$$

a contradiction to Theorem 13.2. This proves Assertion (1).

If (2) would be decidable, then (1) would be decidable, too, since

$$L(\mathcal{G}_1) = L(\mathcal{G}_2) \iff L(\mathcal{G}_1) \subseteq L(\mathcal{G}_2) \text{ and } L(\mathcal{G}_2) \subseteq L(\mathcal{G}_1) , \qquad (14.4)$$

a contradiction. Therefore, we obtain that (2) is not decidable and Assertion (2) is shown. ∎

14.2 Back to Regular Languages

This is a good place to summarize our knowledge about regular languages and to compare the results obtained to the undecidability results for context-free languages shown above.

By Theorem 3.1, we know that $L \in \mathcal{REG}$ if and only if there is a DFA $\mathcal{A}$ such that $L = L(\mathcal{A})$. Thus, we directly obtain the following corollary.

Corollary 14.2 *The regular languages are closed under complement.*

Proof. Let $L \in \mathcal{REG}$ be any language. By Theorem 3.1 there is a DFA $\mathcal{A} = [\Sigma, Q, \delta, q_0, F]$ such that $L = L(\mathcal{A})$. Let $\overline{\mathcal{A}} = [\Sigma, Q, \delta, q_0, Q \setminus F]$. Then, we obviously have $\overline{L} = L(\overline{\mathcal{A}})$. ∎

Furthermore, by Theorem 2.1, the regular languages are closed under union, product and Kleene closure. So, recalling a bit set theory, we know that $L_1 \cap L_2 = \overline{\overline{L_1} \cup \overline{L_2}}$. Hence we directly get the following corollary.

Corollary 14.3 *The regular languages are closed under intersection.*

Moreover, there is an algorithm which on input any regular grammar $\mathcal{G}$ decides whether or not $L(\mathcal{G})$ is infinite or finite (cf. Theorem 4.3).

Additionally, the proof of Theorem 4.3 allows for the following corollary.

Corollary 14.4 *There is an algorithm which on input any regular grammar $\mathcal{G}$ decides whether or not $L(\mathcal{G}) = \emptyset$.*

Proof. Let $\mathcal{G}$ be a regular grammar. The algorithm constructs a DFA $\mathcal{A} = [\Sigma, Q, \delta, q_0, F]$ such that $L(\mathcal{G}) = L(\mathcal{A})$. Let $\text{card}(Q) = n$. Then, the algorithm checks whether or not there is a string s such that

$$n + 1 \leq |s| \leq 2n + 2 \quad \text{and} \quad s \in L(\mathcal{A}) \; .$$

If there is no such string, then by the proof of Theorem 4.3 we already know that $L(\mathcal{G})$ is finite. Thus, it suffices to check whether or not there is a string s such that $|s| \leq n$ and $s \in L(\mathcal{A})$.

If there is such a string, output $L(\mathcal{G}) \neq \emptyset$. Otherwise, output $L(\mathcal{G}) = \emptyset$. ∎

Of course, this is not the most efficient algorithm.

Finally, we have shown that $\mathcal{REG}$ is closed under set difference (cf. Problem 4.2). Thus, we also have the following corollary.

Corollary 14.5 *There is an algorithm which on input any regular grammars $\mathcal{G}_1$ and $\mathcal{G}_2$ decides whether or not $L(\mathcal{G}_1) \subseteq L(\mathcal{G}_2)$.*

Proof. We know that $L(\mathcal{G}_1) \setminus L(\mathcal{G}_2) \in \mathcal{REG}$, since $\mathcal{REG}$ is closed under set difference. Furthermore, a closer inspection of the solution of Problem 4.2 shows that we can construct a grammar $\mathcal{G}$ such that $L(\mathcal{G}) = L(\mathcal{G}_1) \setminus L(\mathcal{G}_2)$. Again recalling a bit set theory, we have $L(\mathcal{G}_1) \subseteq L(\mathcal{G}_2)$ if and only if $L(\mathcal{G}) = \emptyset$. By Corollary 14.4, it is decidable whether or not $L(\mathcal{G}) = \emptyset$. ∎

Using the same idea as in the proof of Corollary 14.1 (cf. (14.4)), we also see that there is an algorithm which on input any regular grammars $\mathcal{G}_1$ and $\mathcal{G}_2$ decides whether or not $L(\mathcal{G}_1) = L(\mathcal{G}_2)$.

Thus, we can conclude that all the problems considered in Theorems 14.1 and 14.2 and Corollary 14.1 when stated *mutatis mutandis* for regular languages are *decidable*.

14.3 Results concerning $\mathcal{L}_0$

We need some more notations which we introduce below.

Let $A \subseteq \mathbb{N}$ be any set. We have already defined the characteristic function χ_A of A. Knowing $\chi_A \in \mathcal{R}$ provides us with full algorithmic control of the set A. But we have already seen that there are sets A for which $\chi_A \notin \mathcal{R}$ holds, e.g., the halting set K. Thus, it is natural to ask whether or not we can weaken the notion of characteristic function in way such

14. Applications of PCP

that some algorithmic control of the set A remains. For establishing the affirmative answer, we introduce the *partial characteristic function* (部分特性関数) $\pi_A : A \to \{1\}$ defined as follows.

$$\pi_A(x) = \begin{cases} 1, & \text{if } x \in A \text{ ;} \\ \text{not defined}, & \text{otherwise .} \end{cases}$$

Now we can formally define what does it mean that a set is recursively enumerable.

Definition 14.1 *Let* $A \subseteq \mathbb{N}$; *then* A *is said to be* recursively enumerable (帰納可算的) *if* $\pi_A \in \mathcal{P}$.

Comparing Definition 14.1 and Definition 13.1, we see that the difference is $\pi_A \in \mathcal{P}$ versus $\chi_A \in \mathcal{R}$.

We continue with some examples for sets being recursive and recursively enumerable, respectively.

(1) The set of all prime numbers is recursive.

(2) The set of all odd numbers is recursive.

(3) The set of all even numbers is recursive.

(4) $\emptyset$ is recursive.

(5) Every finite set is recursive.

(6) Every recursive set is recursively enumerable.

(7) Let A be any recursive set. Then its complement $\overline{A}$ is recursive, too.

Exercise 14.1 *Verify Assertions* (1) *through* (7) *made above.*

The following theorem establishes a characterization of recursive sets.

Theorem 14.3 *Let* $A \subseteq \mathbb{N}$; *then we have:* A *is recursive if and only if* A *is recursively enumerable and* $\overline{A}$ *is recursively enumerable.*

Proof. Necessity. If A is recursive then A is recursively enumerable. Furthermore, as mentioned above, the complement $\overline{A}$ of every recursive set A is recursive, too. Thus, $\overline{A}$ is also recursively enumerable.

Sufficiency. If both A and $\overline{A}$ are recursively enumerable, then $\pi_A \in \mathcal{P}$ and $\pi_{\overline{A}} \in \mathcal{P}$. Thus, we can express χ_A as follows:

$$\chi_A(x) = \begin{cases} 1, & \text{if } \pi_A(x) = 1 \text{ ;} \\ 0, & \text{if } \pi_{\overline{A}}(x) = 1 \text{ .} \end{cases}$$

Consequently, $\chi_A \in \mathcal{R}$. ∎

The latter theorem can be used to show that there are recursively enumerable sets which are not recursive.

Corollary 14.6 *The halting set* K *is recursively enumerable but not recursive.*

Proof. Recall that $K = \{i \mid i \in \mathbb{N}, \psi_i(i) \downarrow\}$, where $\psi \in \mathcal{P}^2$ is universal for $\mathcal{P}$. Since $\psi \in \mathcal{P}^2$, there must exist a TM M such that $f_M^2 = \psi$.

Hence, we have the following equivalence:

$$\pi_K(i) = 1 \iff \psi_i(i) \downarrow \iff M(i,i) \text{ stops .}$$

Thus, in order to compute the function $\pi_K(i)$, it suffices to start the TM M on input (i,i). If the computation terminates, the algorithm for computing $\pi_K(i)$ outputs 1. Otherwise, the algorithm does not terminate. Consequently, $\pi_K \in \mathcal{P}$ and thus K is recursively enumerable.

Suppose K to be recursive. Then we have $\chi_K \in \mathcal{R}$, a contradiction to Theorem 13.1. ∎

By Theorem 14.3, we directly arrive at the following corollary.

Corollary 14.7 *The complement* $\overline{K}$ *of the halting set* K *is not recursively enumerable.*

So, it remains to relate the language family $\mathcal{L}_0$ to Turing machines. In the following, by "any type-0 grammar" we always mean a grammar as defined in Definition 2.1 without any restrictions.

Theorem 14.4 *Let $\mathcal{G}$ be any type-0 grammar. Then there exists a Turing machine M such that $L(\mathcal{G}) = L(M)$.*

The opposite is also true.

Theorem 14.5 *For every Turing machine M there exists a type-0 grammar $\mathcal{G}$ such that $L(M) = L(\mathcal{G})$.*

Theorems 14.4 and 14.5 directly allow for the following corollary.

Corollary 14.8 *$\mathcal{L}_0$ is equal to the family of all recursively enumerable sets.*

Now, we are ready to show the theorem already announced at the end of Chapter 10, i.e., that Theorem 10.9 does not generalize to $\mathcal{L}_0$.

Theorem 14.6 *There does not exist any algorithm that on input any type-0 grammar $\mathcal{G} = [T, N, \sigma, P]$ and any string $s \in T^*$ decides whether or not $s \in L(\mathcal{G})$.*

Proof. By Corollary 14.6, we know that K is recursively enumerable but not recursive. Since $\pi_K \in \mathcal{P}$, there exists a Turing machine computing π_K. So, by Theorem 14.5, there is a grammar $\mathcal{G}$ such that $L(\mathcal{G}) = K$. Thus, if we could decide $s \in L(\mathcal{G})$ for every $s \in \mathbb{N}$, then K would be recursive, a contradiction. ∎

Since $\overline{K}$ is not recursively enumerable (cf. Corollary 14.7), we also have the following corollary.

Corollary 14.9 *$\mathcal{L}_0$ is not closed under complement and set difference.*

Interestingly, the ideas used in the proof of Theorem 2.1 directly yield the following theorem.

Theorem 14.7 *The language family $\mathcal{L}_0$ is closed under union, product and Kleene closure.*

Finally, we show further undecidability results to complete the picture for the language family $\mathcal{L}_0$.

Theorem 14.8 *There does not exist any algorithm which, on input any type-0 grammar $\mathcal{G}$ decides whether or not $L(\mathcal{G}) = \emptyset$.*

Proof. Suppose the converse. Let $\mathcal{G} = [T, N, \sigma, P]$ be any grammar and let any string $w \in T^*$ be arbitrarily fixed. Now, we construct a grammar $\tilde{\mathcal{G}} = [T, N \cup \{\tilde{\sigma}, \#\}, \tilde{\sigma}, \tilde{P}]$, where $\tilde{P} = P \cup \{\tilde{\sigma} \to \#\sigma\#, \#w\# \to \lambda\}$.

Note that $\#w\# \to \lambda$ is the only production which can remove the $\#$ symbols. Thus, we get

$$L(\tilde{\mathcal{G}}) = \begin{cases} \{\lambda\}, & \text{if } w \in L(\mathcal{G}) \ ; \\ \emptyset, & \text{otherwise} \ . \end{cases}$$

Consequently, $w \notin L(\mathcal{G})$ if and only if $L(\tilde{\mathcal{G}}) = \emptyset$. Thus, we can decide $L(\tilde{\mathcal{G}}) = \emptyset$ if and only if we can decide $w \notin L(\mathcal{G})$. But the latter problem is undecidable (cf. Theorem 14.6). Since the construction of $\tilde{\mathcal{G}}$ can be done algorithmically, we obtain a contradiction. ∎

Corollary 14.10 *The following problems are undecidable for any type-0 grammars $\mathcal{G}_1$, $\mathcal{G}_2$:*

(1) $L(\mathcal{G}_1) = L(\mathcal{G}_2)$,

(2) $L(\mathcal{G}_1) \subseteq L(\mathcal{G}_2)$.

Proof. Let $\mathcal{G}_\emptyset$ be any type-0 grammar such that $L(\mathcal{G}_\emptyset) = \emptyset$. Then we have for any type-0 grammar $\mathcal{G}$:

$$L(\mathcal{G}) = \emptyset \iff L(\mathcal{G}) \subseteq L(\mathcal{G}_\emptyset) \iff L(\mathcal{G}) = L(\mathcal{G}_\emptyset) \ ,$$

and thus both the inclusion and equivalence problem, respectively, are reduced to the decidability of the emptiness problem. Since the emptiness problem is undecidable by Theorem 14.8, the corollary is shown. ∎

As a general "rule of thumb" we should memorize that every *non-trivial* problem is undecidable for type-0 grammars. Here by non-trivial we mean that there are infinitely many grammars satisfying the problem and infinitely many grammars not satisfying it.

14.4 Summary

Figure 14.1 summarizes many important results. All problems listed should be read as follows:

Does there exist an algorithm which on input any grammar $\mathcal{G}$ and any string s or any grammar $\mathcal{G}$ or any two grammars $\mathcal{G}_1$, $\mathcal{G}_2$, respectively, returns 1 if the property on hand is fulfilled and 0 if it is not fulfilled.

We use + to indicate that the desired algorithm exists and − to indicate that the desired algorithm does *not* exist.

	Problem	$\mathcal{REG}$	$\mathcal{CF}$	$\mathcal{CS}$	Type 0
1)	$s \in L(\mathcal{G})$	+	+	+	−
2)	$L(\mathcal{G}_1) \subseteq L(\mathcal{G}_2)$	+	−	−	−
3)	$L(\mathcal{G}_1) = L(\mathcal{G}_2)$	+	−	−	−
4)	$L(\mathcal{G}) = \emptyset$	+	+	−	−
5)	$L(\mathcal{G})$ finite	+	+	−	−
6)	$L(\mathcal{G})$ infinite	+	+	−	−
7)	$\overline{L(\mathcal{G})} = \emptyset$	+	−	−	−
8)	$\overline{L(\mathcal{G})}$ infinite	+	−	−	−
9)	$\overline{L(\mathcal{G})}$ has the same type as $L(\mathcal{G})$	+	−	+	−
10)	$\overline{L(\mathcal{G})} \in \mathcal{REG}$	+	−	−	−
11)	$\overline{L(\mathcal{G})} \in \mathcal{CF}$	+	−	−	−
12)	$L(\mathcal{G}_1) \cap L(\mathcal{G}_2) = \emptyset$	+	−	−	−
13)	$L(\mathcal{G}_1) \cap L(\mathcal{G}_2)$ finite	+	−	−	−
14)	$L(\mathcal{G}_1) \cap L(\mathcal{G}_2)$ infinite	+	−	−	−
15)	$L(\mathcal{G}) = T^*$	+	−	−	−
16)	$L(\mathcal{G}) \in \mathcal{REG}$	+	−	−	−

Figure 14.1 Decidable and undecidable problems for formal languages.

Note that for context-free grammars $\mathcal{G}$ the decidability of $s \in L(\mathcal{G})$ is a direct consequence of Theorem 10.9, and the decidability of $L(\mathcal{G})$ finite or infinite can be shown *mutatis mutandis* as Theorem 4.3.

14.5 Problem Set 14

Problem 14.1 Prove the following: Let $A \subseteq \mathbb{N}$ be any set. Then, the following conditions are equivalent.

(1) $\pi_A \in \mathcal{P}$.
(2) There is a function $f \in \mathcal{P}$ such that $A = dom(f)$.
(3) There is a function $g \in \mathcal{P}$ such that $A = range(g)$.
(4) There is a function $h \in \mathcal{R}$ such that $A = range(h)$ or $A = \emptyset$.

Problem 14.2 Let $\psi \in \mathcal{P}^2$ be a universal numbering for $\mathcal{P}$. Consider the set $A = \{i \mid \psi_i \in \mathcal{R}\}$. Prove or disprove A to be recursively enumerable.

Problem 14.3 We call a function $f \colon \mathbb{N} \to \mathbb{N}$ strong monotonic if $n < m$ implies $f(n) < f(m)$ for all $n, m \in \mathbb{N}$. Now, let $f \in \mathcal{R}$ be strong monotonic. Prove or disprove the following assertions to be true.

(1) $dom(f)$ is recursive.
(2) $range(f)$ is recursive.

15
Numberings, Complexity

As we have seen in Chapter 12, there are numberings that are universal for $\mathcal{P}$ (cf. Theorem 12.3). In this chapter we study further properties of numberings. Then, we turn our attention to complexity.

For every $\psi \in \mathcal{P}^2$ we set $\mathcal{P}_\psi = \{\psi_i \mid i \in \mathbb{N}\}$. Next, we introduce the notion of reducibility.

Definition 15.1 *Let* $\psi, \psi' \in \mathcal{P}^2$ *be any numberings. We say that* ψ *is* reducible (帰着可能) *to* ψ' *(written* $\psi \leq \psi'$*) if there exists a function* $c \in \mathcal{R}$ *such that* $\psi_i = \psi'_{c(i)}$ *for all* $i \in \mathbb{N}$.

Clearly, if $\psi \leq \psi'$ then $\mathcal{P}_\psi \subseteq \mathcal{P}_{\psi'}$. If we interpret every i as a program, then the function c reducing ψ to ψ' can be interpreted as a *compiler* (コンパイラ). This is the reason we called the reduction function c.

Exercise 15.1 *Prove reducibility to be reflexive and transitive.*

It should be noted that reducibility is not trivial. Friedberg [8] has shown the following theorem (see also Kummer [20] for an easier proof).

Theorem 15.1 *There are numberings* $\psi, \psi' \in \mathcal{P}^2$ *such that* $\mathcal{P}_\psi = \mathcal{P}_{\psi'}$ *but neither* $\psi \leq \psi'$ *nor* $\psi' \leq \psi$.

We do not prove this theorem here, since we do not need it. However, we mention that in Theorem 15.1 $\psi, \psi' \in \mathcal{P}^2$ may be chosen such that $\mathcal{P}_\psi = \mathcal{P}_{\psi'} = \mathcal{P}$, i.e., ψ, ψ' can be universal for $\mathcal{P}$.

Thus, "$\leq$" is a partial order. We ask whether or not there are numberings that are maximal with respect to reducibility. The answer is given below.

15.1 Gödel Numberings

Definition 15.2 *Let $\varphi \in \mathcal{P}^2$ be a numbering. We call φ a Gödel numbering (ゲーデル数化) if*
(1) $\mathcal{P}_\varphi = \mathcal{P}$, *and*
(2) $\psi \leq \varphi$ *for every numbering $\psi \in \mathcal{P}^2$.*

The following theorem establishes the existence of Gödel numberings.

Theorem 15.2 *There exists a Gödel numbering.*

Proof. In fact, we have already constructed a Gödel numbering when proving Theorem 12.3. So, let φ be the numbering corresponding to the universal Turing machine U. Thus, we already know $\mathcal{P}_\varphi = \mathcal{P}$.

Claim 1. For every numbering $\psi \in \mathcal{P}^2$ there is a function $c \in \mathcal{R}$ such that $\psi_i(x) = \varphi_{c(i)}(x)$ for all $i, x \in \mathbb{N}$.

Since $\psi \in \mathcal{P}^2$, there exists a TM M computing ψ. Now, if we fix the first argument i then we get a TM M_i computing ψ_i. Then we set $c(i) = cod(M_i)$, where cod is the computable function from Theorem 12.3. Consequently, $c(i)$ is defined for every $i \in \mathbb{N}$ and $cod(M_i)$ can be computed from the knowledge of M. Finally, we have

$$\psi_i(x) = f_M(i, x) = f_{M_i}(x) = f_U(cod(M_i), x) = f_U(c(i), x) = \varphi_{c(i)}(x),$$

and Claim 1 follows. ∎

Now that we know that there is one Gödel numbering, it is only natural to ask if it is the only one. The answer is provided by the next theorem.

Theorem 15.3 *There are countably many Gödel numberings for $\mathcal{P}$.*

Proof. Let φ be the Gödel numbering from Theorem 15.2 and let $\psi \in \mathcal{P}^2$ be any numbering. We can thus define $\varphi' \in \mathcal{P}^2$ by setting $\varphi'_{2i} = \varphi_i$ and $\varphi'_{2i+1} = \psi_i$ for all $i \in \mathbb{N}$.

Clearly, φ' is universal for $\mathcal{P}$. Moreover, using $c(i) = 2i$ we get $\varphi \leq \varphi'$. Since $\leq$ is transitive, we thus have shown φ' to be a Gödel numbering. ∎

So, the Gödel numberings are the maximal elements of the lattice $[\mathcal{P}, \leq]$. Moreover all Gödel numberings are equivalent in the following sense.

Theorem 15.4 (Rogers' Theorem (ロジャースの定理) **[32]).** *For any two Gödel numberings φ and ψ of $\mathcal{P}$ there exists a permutation $\pi \in \mathcal{R}$ such that for all $i \in \mathbb{N}$ we have $\varphi_i = \psi_{\pi(i)}$ and $\psi_i = \varphi_{\pi^{-1}(i)}$.*

In the following by **Göd** we denote the set of all Gödel numberings for $\mathcal{P}$.

15.2 The Fixed Point, the Recursion, and Rice's Theorems

The following theorem has been discovered by Rogers [33]. It is called the *fixed point theorem* (不動点定理), since it states that every recursive transformation of programs leaves at least the input/output behavior of one program unchanged. Thus, it establishes the existence of at least one semantical fixed point.

Theorem 15.5 (Fixed Point Theorem). *Let $\varphi \in$ Göd. Then, for every function $h \in \mathcal{R}$ there exists a number $a \in \mathbb{N}$ such that $\varphi_{h(a)} = \varphi_a$.*

Proof. First we define a function τ as follows. For all $i, x \in \mathbb{N}$ let

$$\tau(i, x) = \begin{cases} \varphi_{\varphi_i(i)}(x), & \text{if } \varphi_i(i) \downarrow ; \\ \text{not defined}, & \text{otherwise} . \end{cases}$$

Clearly, $\tau \in \mathcal{P}^2$ and thus τ is a numbering, too. By construction we obtain

$$\tau_i = \begin{cases} \varphi_{\varphi_i(i)}, & \text{if } \varphi_i(i) \downarrow ; \\ e, & \text{otherwise} , \end{cases}$$

where e denotes the nowhere defined function. By the definition of a Gödel

15.2 The Fixed Point, the Recursion, and Rice's Theorems

numbering there exists a function $c \in \mathcal{R}$ such that

$$\tau_i = \varphi_{c(i)} \text{ for all } i \in \mathbb{N}. \tag{15.1}$$

Next, we consider $g(x) = h(c(x))$ for all $x \in \mathbb{N}$. Since $h, c \in \mathcal{R}$, we also know that $g \in \mathcal{R}$. Since φ is universal for $\mathcal{P}$, there exists a $v \in \mathbb{N}$ such that

$$g = \varphi_v. \tag{15.2}$$

Consequently, $\varphi_v(v) \downarrow$ because of $g \in \mathcal{R}$. Hence, we obtain

$\tau_v = \varphi_{c(v)}$ by the definition of c, see (15.1)

     $= \varphi_{\varphi_v(v)}$ by the definition of τ and because of $\varphi_v(v) \downarrow$

     $= \varphi_{g(v)}$ since $g = \varphi_v$, see (15.2)

     $= \varphi_{h(c(v))}$ by the definition of g.

Thus, $\varphi_{c(v)} = \varphi_{h(c(v))}$. Hence, setting $a = c(v)$ proves the theorem. ■

The latter proof even shows that fixed points can be computed.

Corollary 15.1 *For every $\varphi \in$ Göd there exists a function fix $\in \mathcal{R}$ such that for all $z \in \mathbb{N}$: If $\varphi_z \in \mathcal{R}$ then $\varphi_{\varphi_z(fix(z))} = \varphi_{fix(z)}$.*

The following theorem is due to Kleene [18]. It is called *recursion theorem* (再帰定理). Here recursion is referring to computational self-reference.

Theorem 15.6 (Recursion Theorem). *For every numbering $\psi \in \mathcal{P}^2$ and every $\varphi \in$ Göd there exists a number $a \in \mathbb{N}$ such that $\varphi_a(x) = \psi(a, x)$ for all $x \in \mathbb{N}$.*

Proof. Since $\psi \in \mathcal{P}^2$ and since $\varphi \in \mathcal{P}^2$ is a Gödel numbering, there exists a function $c \in \mathcal{R}$ such that $\psi_i = \varphi_{c(i)}$ for all $i \in \mathbb{N}$.

By Theorem 15.5 and the choice of function c, there is a number $a \in \mathbb{N}$ such that $\psi_a = \varphi_{c(a)} = \varphi_a$, and thus $\psi(a, x) = \varphi_a(x)$ for all $x \in \mathbb{N}$. ■

Note that the recursion theorem also implies the fixed point theorem. We leave the proof as an exercise.

There are important generalizations of the fixed point and recursion theorem. We mention here one which has been found by Smullyan [35].

Theorem 15.7 (Smullyan's Double Fixed Point Theorem). *Let $\varphi \in$ Göd and let $h_1, h_2 \in \mathcal{R}^2$. Then there exist $i_1, i_2 \in \mathbb{N}$ such that simultaneously $\varphi_{i_1} = \varphi_{h_1(i_1,i_2)}$ and $\varphi_{i_2} = \varphi_{h_2(i_1,i_2)}$ are satisfied.*

Exercise 15.2 *Let $\varphi \in$ Göd. Prove that there always exists an $i \in \mathbb{N}$ such that $\varphi_i = \varphi_{i+1}$.*

Next we show that all nontrivial properties of programs are undecidable. First, we define what is commonly called an index set.

Definition 15.3 *Let $\varphi \in$ Göd and let $\mathcal{P}' \subseteq \mathcal{P}$. Then the set $\Theta_\varphi \mathcal{P}' = \{i \mid i \in \mathbb{N} \text{ and } \varphi_i \in \mathcal{P}'\}$ is called* index set (索引集合) *of $\mathcal{P}'$.*

The term "index" in the definition above is synonymous in its use to the term "program." However, we follow here the traditional terminology, since *program set* could lead to confusion. One may be tempted to interpret *program set* as set of programs, i.e., a collection of programs that may or may not be an index set.

Theorem 15.8 (Rice's Theorem (ライスの定理) **[31])** *Let $\varphi \in$ Göd. Then $\Theta_\varphi \mathcal{P}'$ is undecidable for every set $\mathcal{P}'$ with $\emptyset \subset \mathcal{P}' \subset \mathcal{P}$.*

Proof. Suppose the converse, i.e., there is a set $\mathcal{P}'$ such that $\emptyset \subset \mathcal{P}' \subset \mathcal{P}$ and $\Theta_\varphi \mathcal{P}'$ is decidable. Thus $\chi_{\Theta_\varphi \mathcal{P}'} \in \mathcal{R}$. Since $\emptyset \subset \mathcal{P}' \subset \mathcal{P}$, there is a function $g \in \mathcal{P} \setminus \mathcal{P}'$ and a function $f \in \mathcal{P}'$. Let u be any program for g, i.e., $\varphi_u = g$ and let z be any program for f, i.e., $\varphi_z = f$. Next, we define a function h as follows. For all $i \in \mathbb{N}$ we set

$$h(i) = \begin{cases} u, & \text{if } \chi_{\Theta_\varphi \mathcal{P}'}(i) = 1 \text{ ;} \\ z, & \text{otherwise .} \end{cases}$$

Since $\chi_{\Theta_\varphi \mathcal{P}'} \in \mathcal{R}$, we conclude that $h \in \mathcal{R}$. Hence, by Theorem 15.5, there is an $a \in \mathbb{N}$ such that $\varphi_{h(a)} = \varphi_a$. We distinguish the following cases.

15.2 The Fixed Point, the Recursion, and Rice's Theorems

Case 1. $\varphi_a \in \mathcal{P}'$.

Then $\chi_{\Theta_\varphi \mathcal{P}'}(a) = 1$, and thus $h(a) = u$. Consequently,

$$\varphi_a = \varphi_{h(a)} = \varphi_u = g \notin \mathcal{P}', \quad \text{a contradiction to } \chi_{\Theta_\varphi \mathcal{P}'}(a) = 1 \,.$$

Case 2. $\varphi_a \notin \mathcal{P}'$.

Then $\chi_{\Theta_\varphi \mathcal{P}'}(a) = 0$, and therefore $h(a) = z$. Consequently,

$$\varphi_a = \varphi_{h(a)} = \varphi_z = f \in \mathcal{P}', \quad \text{a contradiction to } \chi_{\Theta_\varphi \mathcal{P}'}(a) = 0 \,.$$

Therefore, the set $\Theta_\varphi \mathcal{P}'$ is undecidable. ∎

Example 15.1 Let $\mathcal{P}' = Prim$. As we have seen, $Prim \neq \emptyset$. By Theorem 11.4 we also have $Prim \subset \mathcal{P}$. Consequently, $\Theta_\varphi Prim$ is undecidable. That is, there does not exist any algorithm which on input any program $i \in \mathbb{N}$ can decide whether or not it computes a primitive recursive function.

Rice's theorem directly allows for the following corollary.

Corollary 15.2 *Let* $\varphi \in$ *Göd. Then for every function* $f \in \mathcal{P}$ *the set* $\{i \mid i \in \mathbb{N} \text{ and } \varphi_i = f\}$ *is infinite.*

Proof. Setting $\mathcal{P}' = \{f\}$, we get $\emptyset \subset \mathcal{P}' \subset \mathcal{P}$. Thus the assumptions of Rice's theorem are fulfilled. Therefore $\Theta_\varphi \mathcal{P}'$ is undecidable. On the other hand, $\Theta_\varphi \mathcal{P}' = \{i \mid i \in \mathbb{N} \text{ and } \varphi_i = f\}$. Since every finite set is decidable, we can conclude that $\{i \mid i \in \mathbb{N} \text{ and } \varphi_i = f\}$ is infinite. ∎

Exercise 15.3 *Let* $\varphi \in$ *Göd. Prove or disprove the following sets to be recursive.*

(1) *Let* $e: \mathbb{N} \to \mathbb{N}$ *be the nowhere defined function. Consider the set* $A = \{i \mid dom(\varphi_i) = dom(e)\}$.

(2) *Consider the set* $B = \{i \mid range(\varphi_i) = \{1\}\}$.

(3) *Consider the set* $C = \{(i, j) \mid \varphi_i = \varphi_j\}$.

15.3 Complexity

Next, we turn our attention to complexity. Using the setting of recursive functions, one can show several results that turn out to be very useful and insightful. So, we shall define abstract complexity measures and prove some fundamental results.

We make the following conventions. The quantifiers "$\forall^\infty$" and "$\exists^\infty$" are interpreted as "for all but finitely many" and "there exists infinitely many," respectively. For any set $S \subseteq \mathbb{N}$, by $\max S$ and $\min S$ we denote the maximum and minimum of a set S, respectively, where, by convention, $\max \emptyset = 0$ and $\min \emptyset = \infty$.

Now, we are ready to define (abstract) complexity measures.

Definition 15.4 (Blum [2]). *Let* $\varphi \in \text{Göd}$ *and let* $\Phi \in \mathcal{P}^2$. *We call* $[\varphi, \Phi]$ complexity measure (計算量尺度) *if*

(1) $dom(\varphi_i) = dom(\Phi_i)$ *for all* $i \in \mathbb{N}$, *and*

(2) *there exist a recursive predicate* M *such that*

$$\forall i \forall n \forall y [M(i, n, y) = 1 \iff \Phi_i(n) = y] \ .$$

Note that Condition (1) and (2) are independent from one another. This can be seen as follows. Let φ be any Gödel numbering and define $\Phi_i(n) = 0$ for all $i, n \in \mathbb{N}$. Then $\Phi \in \mathcal{P}^2$ (in fact $\Phi \in \mathcal{R}^2$) and it obviously satisfies Condition (2). Of course, Condition (1) is violated.

Next consider

$$\Phi_i(n) = \begin{cases} 0, & \text{if } \varphi_i(n) \text{ is defined;} \\ \text{not defined}, & \text{otherwise.} \end{cases}$$

Now, Condition (1) is obviously fulfilled but Condition (2) is not satisfied. A well-known example of a complexity measure is the number of steps required by the ith TM (in a standard enumeration of all TMs) to converge on input x, i.e., the standard *time measure* (時間尺度). Another example is the *space measure* (空間尺度), i.e., the number of cells containing non-blank symbols or visited by the read-write head during a computation by a TM, provided the latter is considered undefined if the machine loops on a bounded tape segment. Further complexity measures comprise

(a) *reversal*, i.e., the number of times during the computation of the ith TM that the head must change direction.

(b) *ink*, i.e., the number of times during the computation of the ith TM that a symbol has to be overwritten by a different symbol.

Exercise 15.4 *Prove or disprove that* carbon, *i.e., the number of times during the computation of the ith TM that a symbol has to be overwritten by the same symbol is a complexity measure.*

Next, we establish several basic properties. The following theorem shows that the function values of all partial recursive functions are uniformly boundable by their complexities. But before we can state the result formally, we have to explain what does it mean that $\psi(x) = \theta(x)$ for partial functions ψ and θ and $x \in \mathbb{N}$. By $\psi(x) = \theta(x)$ we mean that either both values $\psi(x)$ and $\theta(x)$ are defined and $\psi(x) = \theta(x)$ or else both values $\psi(x)$ and $\theta(x)$ are undefined. Analogously, we define $\psi(x) \leq \theta(x)$.

Theorem 15.9 *Let $[\varphi, \Phi]$ be a complexity measure. Then there exists a function $h \in \mathcal{R}^2$ such that for all $i \in \mathbb{N}$ and all but finitely many $n \in \mathbb{N}$ we have $\varphi_i(n) \leq h(n, \Phi_i(n))$.*

Proof. We define the desired function h as follows. For all $n, t \in \mathbb{N}$, let

$$h(n, t) = \max\{\varphi_i(n) \mid i \leq n \text{ and } \Phi_i(n) = t\} .$$

Since the predicate "$\Phi_i(n) = t$" is recursive, we see that $h \in \mathcal{R}^2$.

It remains to show that h fulfills the desired property. First, suppose that $\varphi_i(n) \uparrow$. Then, by Condition (1) of Definition 15.4, we also know that $\Phi_i(n) \uparrow$. Hence, the value $h(n, \Phi_i(n))$ is undefined, too, and thus the inequality $\varphi_i(n) \leq h(n, \Phi_i(n))$ is satisfied.

Next, assume $\varphi_i(n) \downarrow$. By the same argument as above we conclude that $\Phi_i(n) \downarrow$. Thus $h(n, \Phi_i(n))$ is defined. Moreover, if $i \leq n$ then, by construction, we have $\varphi_i(n) \leq h(n, \Phi_i(n))$. Since there are only finitely many $n < i$, we are done. ∎

Intuitively speaking, Theorem 15.9 says that rapidly growing functions must be also "very" complex. This is of course obvious for a measure like time, since it takes a huge amount of time to write very large numbers down. It is, however, by no means obvious for complexity measures like reversal. So, maybe there is a deeper reason for this phenomenon. This is indeed the case, since all complexity measures can be recursively related to one another as our next theorem shows (cf. Blum [2]).

Theorem 15.10 (Blum [2]) *Let $[\varphi, \Phi]$ and $[\psi, \Psi]$ be any two complexity measures. Let π be a recursive permutation such that $\varphi_i = \psi_{\pi(i)}$ for all $i \in \mathbb{N}$. Then there exists a function $h \in \mathcal{R}^2$ such that*

$$\forall i \forall^\infty n \, [\, \Phi_i(n) \leq h(n, \Psi_{\pi(i)}(n)) \quad \text{and} \quad \Psi_{\pi(i)}(n) \leq h(n, \Phi_i(n)) \,] \, .$$

Proof. We define the desired function h as follows. For all $n, t \in \mathbb{N}$ let

$$h(n, t) = \max\{\Phi_i(n) + \Psi_{\pi(i)}(n) \mid i \leq n \wedge (\Phi_i(n) = t \vee \Psi_{\pi(i)}(n) = t)\}.$$

We show that h fulfills the properties stated. Since $[\varphi, \Phi]$ and $[\psi, \Psi]$ are complexity measures, the predicates "$\Phi_i(n) = t$" and "$\Psi_{\pi(i)}(n) = t$" are both recursive in i, n, t. Moreover, by Condition (2) of Definition 15.4, if $\Phi_i(n) = t$ then, in particular, $\varphi_i(n) \downarrow$. Because of $\varphi_i = \psi_{\pi(i)}$, we conclude that $\psi_{\pi(i)}(n) \downarrow$, too. Now another application of Condition (2)

15.3 Complexity

of Definition 15.4 directly yields that also $\Psi_{\pi(i)}(n) \downarrow$. Analogously it can be shown that $\Psi_{\pi(i)}(n) = t$ implies that $\Phi_i(n) \downarrow$. Thus, if $\Phi_i(n) = t$ or $\Psi_{\pi(i)}(n) = t$ for some $i \leq n$, then $h(n,t)$ is defined. If neither $\Phi_i(n) = t$ nor $\Psi_{\pi(i)}(n) = t$ for all $i \leq n$, then we take the maximum of the empty set, i.e., in this case we have $h(n,t) = 0$. Hence, we have $h \in \mathcal{R}^2$. Finally,

$$\forall i \forall^\infty n \ [\ \Phi_i(n) \leq h(n, \Psi_{\pi(i)}(n)) \ \text{ and } \ \Psi_{\pi(i)}(n) \leq h(n, \Phi_i(n)) \ ]$$

follows directly from our construction. We omit the details. ∎

The following theorem establishes a fundamental basic result, i.e., we show that *no recursive amount* of computational resources is sufficient to compute *all* recursive functions. The proof below uses three basic proof techniques: *finite extension* (有限拡大), *diagonalization* (対角化) and *cancellation* (無効化). The idea of finite extension is to construct a function by defining a finite piece at a time. Diagonalization is used to ensure that all functions not fulfilling a certain property must differ from our target function. And cancellation is a technique to keep track of all those programs we have already diagonalized against, and which ones we have yet to consider. Once a program i is diagonalized against, we shall cancel it. Canceled programs are never considered later for future diagonalization. Below, we use $\mathcal{R}_{0,1}$ to denote the set of all $f \in \mathcal{R}$ with $range(f) \subseteq \{0,1\}$.

Theorem 15.11 (Blum [2]) *Let $[\varphi, \Phi]$ be any complexity measure. For every function $h \in \mathcal{R}$ there is a function $f \in \mathcal{R}_{0,1}$ such that for all programs i with $\varphi_i = f$ we have $\Phi_i(n) > h(n)$ for all but finitely many $n \in \mathbb{N}$.*

Proof. We use $\oplus$ as a symbol for addition modulo 2. We define sets C_n in which we keep track of the programs already canceled. The desired function f is defined as follows.

$$f(0) = \begin{cases} \varphi_0(0) \oplus 1, & \text{if } \Phi_0(0) \leq h(0), \text{ then set } C_0 = \{0\} \ ; \\ 0, & \text{otherwise, then set } C_0 = \emptyset \ . \end{cases}$$

Now suppose, $f(0), \ldots, f(n)$ are already defined. We set

$$f(n+1) = \begin{cases} \varphi_{i^*}(n+1) \oplus 1, & \text{if } i^* = \mu i[i \leq n+1,\ i \notin C_n, \\ & \Phi_i(n+1) \leq h(n+1)] \text{ exists}, \\ & \text{then set } C_{n+1} = C_n \cup \{i^*\}\,; \\ 0, & \text{otherwise, then set } C_{n+1} = C_n\,. \end{cases}$$

It remains to show that f satisfies the stated properties. Since the predicate "$\Phi_i(n) = y$" is recursive, so is the predicate "$\Phi_i(n) \leq y$." Now, since $h \in \mathcal{R}$, we can effectively test whether or not $\Phi_0(0) \leq h(0)$. If it is, by Condition (1) of Definition 15.4 we can conclude that $\varphi_0(0) \downarrow$. Hence, in this case $f(0) \downarrow$ and $f(0)$ takes a value from $\{0, 1\}$. Otherwise, $f(0) = 0$, and thus again $f(0) \in \{0, 1\}$. Consequently, our initialization is recursive.

Using the same arguments as above and noting that we only have to check finitely many programs, it is clear that the induction step is recursive, too. By construction, $f(n+1) \in \{0, 1\}$ and so $f \in \mathcal{R}_{0,1}$.

Suppose there is a program i such that $\varphi_i = f$ and $\exists^\infty n\ [\Phi_i(n) \leq h(n)]$. We set $C = \bigcup_{n \in \mathbb{N}} C_n$. By construction, it is easy to see that $i \notin C$, since otherwise $\varphi_i \neq f$ should hold.

Next, we consider $C^{(i)} = \{j \mid j < i \land j \in C\}$. Then we obtain $C^{(i)} \subseteq C$ and $\text{card}(C^{(i)})$ is finite. Therefore, there must be an $m \geq i$ such that $C^{(i)} \subseteq C_n$ for all $n \geq m$. Since there are infinitely many $n \in \mathbb{N}$ with $\Phi_i(n) \leq h(n)$, there is an $n^* \geq m$ such that $\Phi_i(n^*) \leq h(n^*)$. But now

$$i = \mu j[j \leq n^* \land j \notin C_{n^*-1} \land \Phi_i(n^*) \leq h(n^*)]\,.$$

Hence, we have to cancel i when constructing $f(n^*)$, and thus $i \in C_{n^*} \subseteq C$, a contradiction. Consequently, for all but finitely many $n \in \mathbb{N}$ the condition $\Phi_i(n) > h(n)$ must hold. Since i was any program for f, we are done. ∎

Theorem 15.9 showed that we can recursively bound the function values of all partial recursive functions by their complexities. Thus, it is only natural to ask whether or not we can do the converse, i.e., bound the complexity values of all partial recursive functions by their function values. The negative answer is provided next.

Theorem 15.12 *Let $[\varphi, \Phi]$ be any complexity measure. Then there is no recursive function $h \in \mathcal{R}^2$ such that $\forall i \forall^\infty n[\Phi_i(n) \leq h(n, \varphi_i(n))]$.*

Proof. Suppose the converse, i.e., there is a function $h \in \mathcal{R}^2$ such that $\forall i \forall^\infty n[\Phi_i(n) \leq h(n, \varphi_i(n))]$. Consider the recursive function h^* defined as $h^*(n) = h(n, 0) + h(n, 1)$ for all $n \in \mathbb{N}$. By Theorem 15.11 there is a function $f \in \mathcal{R}_{0,1}$ such that $\forall i[\varphi_i = f \implies \forall^\infty n[\Phi_i(n) > h^*(n)]]$. Hence,

$$\forall^\infty n[\Phi_i(n) > h^*(n) = h(n, 0) + h(n, 1) \geq h(n, \varphi_i(n))]$$

for every program i with $\varphi_i = f$, a contradiction to the choice of h. ∎

Next, we ask a slight modification of the question posed before Theorem 15.12, i.e., are there functions $h \in \mathcal{R}^2$ such that there are also functions $f \in \mathcal{P}$ satisfying

$$\exists i[\varphi_i = f \wedge \forall^\infty n[\Phi_i(n) \leq h(n, \varphi_i(n))]] \quad ? \tag{15.3}$$

Before providing the affirmative answer, we make the following definition.

Definition 15.5 *Let $[\varphi, \Phi]$ be a complexity measure, let $h \in \mathcal{R}^2$ and let $\psi \in \mathcal{P}$. The function ψ is said to be h-honest (h-正当) if*

$$\exists i[\varphi_i = \psi \wedge \forall^\infty n[\Phi_i(n) \leq h(n, \varphi_i(n))]] \ .$$

By $H(h)$ we denote the set of all h-honest functions from $\mathcal{P}$.

Now, we are ready to answer the question from (15.3).

Theorem 15.13 *Let $[\varphi, \Phi]$ be a complexity measure and let $\Psi \in \mathcal{P}^2$ be any function such that the predicate "$\Psi_i(x) = y$" is recursive. Then there exists a function $h \in \mathcal{R}^2$ such that $\Psi_i \in H(h)$ for all $i \in \mathbb{N}$.*

Proof. Since $\varphi \in \mathbf{G\ddot{o}d}$ and $\Psi \in \mathcal{P}^2$, there exists a function $c \in \mathcal{R}$ such that $\Psi_i = \varphi_{c(i)}$ for all $i \in \mathbb{N}$. By assumption the predicate "$\Psi_i(x) = y$" is recursive. Therefore, we can define a recursive function $h \in \mathcal{R}^2$ as follows:

$$h(n,t) = \max\{\Phi_{c(i)}(n) \mid i \leq n \wedge \Psi_i(n) = t\}.$$

Consider any Ψ_i. Then for the φ-program $c(i)$ of Ψ_i we have

$$\Psi_i = \varphi_{c(i)} \wedge \forall^\infty n[\Phi_{c(i)}(n) \leq h(n, \varphi_{c(i)}(n))]$$

by construction. ∎

The latter theorem immediately allows for the following corollary.

Corollary 15.3 *Let $[\varphi, \Phi]$ be a complexity measure. Then there exists a function $h \in \mathcal{R}^2$ such that $\{\Phi_i \mid i \in \mathbb{N}\} \subseteq H(h)$.*

Furthermore, Corollary 15.3 and Theorem 15.12 together directly imply the following set theoretical properties of $S = \{\Phi_i \mid i \in \mathbb{N}\}$.

Corollary 15.4 *Let $[\varphi, \Phi]$ be a complexity measure. Then we have*

(1) $S \subset \mathcal{P}$,

(2) $S \cap (\mathcal{P} \setminus \mathcal{R}) \subset \mathcal{P} \setminus \mathcal{R}$, *and*

(3) $S \cap \mathcal{R} \subset \mathcal{R}$.

Finally, it should be noted that it is not meaningful to consider the following stronger version of h-honest functions ψ

$$\forall i[\varphi_i = \psi \wedge \forall^\infty n[\Phi_i(n) \leq h(n, \varphi_i(n))]].$$

This is due to the fact, well-known to everybody who has already written a couple of programs, that there are arbitrarily bad ways of computing any function with respect to any complexity measure. More formally, we have the following theorem (cf. Blum [2]).

Theorem 15.14 *Let $[\varphi, \Phi]$ be any complexity measure. Then, for every program i and every recursive function $h \in \mathcal{R}$ there exists a program b such that $\varphi_b = \varphi_i$ and $\Phi_b(x) > h(x)$ for all $x \in dom(\varphi_i)$.*

Proof. Let $g \in \mathcal{R}$ be any function such that for all $i, n \in \mathbb{N}$

$$\varphi_{g(i)}(n) = \begin{cases} \varphi_i(n), & \text{if } \neg[\Phi_i(n) \leq h(n)] \; ; \\ \varphi_i(n) + 1, & \text{if } \Phi_i(n) \leq h(n) \; . \end{cases}$$

By Theorem 15.5, there is a b such that $\varphi_{g(b)} = \varphi_b$. It remains to show that $\varphi_b = \varphi_i$ and $\Phi_b(x) > h(x)$ for all $x \in dom(\varphi_i)$.

Suppose there is an $x \in dom(\varphi_i)$ such that $\Phi_b(x) \leq h(x)$. By Condition (1) of Definition 15.4 we conclude that $\varphi_b(x) \downarrow$. Since $\varphi_{g(b)} = \varphi_b$ and because of $\Phi_b(x) \leq h(x)$, we then have $\varphi_b(x) = \varphi_{g(b)}(x) = \varphi_b(x) + 1$. Thus, the case $\Phi_b(x) \leq h(x)$ cannot happen. So for all $x \in dom(\varphi_i)$ the case $\Phi_b(x) > h(x)$ must occur, and thus, by construction, $\varphi_b(x) = \varphi_i(x)$ for all $x \in dom(\varphi_i)$. If $x \notin dom(\varphi_i)$, it also holds $\neg[\Phi_b(x) \leq h(x)]$ and so again $\varphi_b(x) = \varphi_i(x)$. This implies $\varphi_i = \varphi_b$ and completes the proof. ∎

Now, we turn our attention to some more advanced topics. We start with complexity classes.

15.4 Complexity Classes

Probably, we have already heard a bit about some complexity classes such as the class of all problems decidable in deterministic polynomial time, the class of all problems decidable in non-deterministic polynomial time, the class of all languages acceptable in deterministic logarithmic space, the class of all languages acceptable in non-deterministic logarithmic space, and the class of all languages acceptable in polynomial space.

For the time being, we want to take a more abstract view of complexity classes such as the ones mentioned above and study properties they have or do not have in common. Therefore, we continue by defining abstract complexity classes (cf. McCreight and Meyer [25]).

Definition 15.6 *Let $[\varphi, \Phi]$ be a complexity measure and $t \in \mathcal{R}$. We set*

$$\mathcal{C}_t^{[\varphi,\Phi]} = \{f \mid f \in \mathcal{R} \wedge \exists i[\varphi_i = f \wedge \forall^\infty n\, \Phi_i(n) \leq t(n)]\},$$

and call $\mathcal{C}_t^{[\varphi,\Phi]}$ *the complexity class (計算量クラス) generated by t.*

First, we deal with the structure of the classes $\mathcal{C}_t^{[\varphi,\Phi]}$. For this purpose, let us recall the common two definitions of enumerability.

Definition 15.7 *Let $\mathcal{U} \subseteq \mathcal{R}$. We call $\mathcal{U}$ recursively enumerable if there exists a function $g \in \mathcal{R}^2$ such that $\mathcal{U} \subseteq \{\lambda x.g(i,x) \mid i \in \mathbb{N}\}$.*

By NUM we denote the collection of all classes $\mathcal{U} \subseteq \mathcal{R}$ that are recursively enumerable.

Definition 15.8 *Let $\mathcal{U} \subseteq \mathcal{R}$. We call $\mathcal{U}$ sharply recursively enumerable if there exists a function $g \in \mathcal{R}^2$ such that $\mathcal{U} = \{\lambda x.g(i,x) \mid i \in \mathbb{N}\}$.*

By NUM! we denote the collection of all classes $\mathcal{U} \subseteq \mathcal{R}$ that are sharply recursively enumerable.

Exercise 15.5 *Prove the following definitions to be equivalent to Definition 15.7 and Definition 15.8, respectively. Let $\varphi \in$ Göd; then we have:*

(1) $\mathcal{U} \in$ NUM *if and only if there exists a function $h \in \mathcal{R}$ such that* $\mathcal{U} \subseteq \{\varphi_{h(i)} \mid i \in \mathbb{N}\} \subseteq \mathcal{R}$.

(2) $\mathcal{U} \in$ NUM! *if and only if there exists a function $h \in \mathcal{R}$ such that* $\mathcal{U} = \{\varphi_{h(i)} \mid i \in \mathbb{N}\} \subseteq \mathcal{R}$.

The following theorem shows in particular that every class $\mathcal{U} \in$ NUM can be embedded into a complexity class.

Theorem 15.15 *Let $[\varphi, \Phi]$ be a complexity measure and let $\mathcal{U} \subseteq \mathcal{R}$ be any class of recursive functions such that $\mathcal{U} \in$ NUM. Then, we have:*

(1) *There exists a function $b \in \mathcal{R}$ such that for all $f \in \mathcal{U}$ the condition $f(n) \leq b(n)$ for all but finitely many $n \in \mathbb{N}$ is satisfied.*

(2) *There exists a function $t \in \mathcal{R}$ such that $\mathcal{U} \subseteq \mathcal{C}_t^{[\varphi,\Phi]}$.*

Proof. There is a function $g \in \mathcal{R}^2$ such that $\mathcal{U} \subseteq \{\lambda x g(i,x) \mid i \in \mathbb{N}\}$, since

$U \in \text{NUM}$. We define the desired function b as follows:

$$b(n) = \max\{g(i,n) \mid i \leq n\} \quad \text{for all } n \in \mathbb{N}.$$

Then function b obviously satisfies the properties stated in (1).

To show Condition (2) we apply Exercise 15.5. So, there is a function $h \in \mathcal{R}$ such that $U \subseteq \{\varphi_{h(i)} \mid i \in \mathbb{N}\} \subseteq \mathcal{R}$. Hence, for all $n \in \mathbb{N}$ we define

$$t(n) = \max\{\Phi_{h(i)}(n) \mid i \leq n\}.$$

The choice of h implies $\varphi_{h(i)} \in \mathcal{R}$ for all $i \in \mathbb{N}$. Using (1) of Definition 15.4, we get $\Phi_{h(i)} \in \mathcal{R}$ for all $i \in \mathbb{N}$. Thus, $t \in \mathcal{R}$ and by construction

$$\forall i \forall^\infty n [\Phi_{h(i)}(n) \leq t(n)].$$

Therefore, we have shown that $U \subseteq \mathcal{C}_t^{[\varphi,\Phi]}$. ∎

15.4.1 Recursive Enumerability of Complexity Classes

Let $[\varphi, \Phi]$ be any complexity measure. We are interested in learning for which $t \in \mathcal{R}$ the complexity class $\mathcal{C}_t^{[\varphi,\Phi]}$ is recursively enumerable. This would be a nice property, since $\mathcal{C}_t^{[\varphi,\Phi]} \in \text{NUM}$ would give us an effective overview about at least one program for every function $f \in \mathcal{C}_t^{[\varphi,\Phi]}$. We show that $\mathcal{C}_t^{[\varphi,\Phi]} \in \text{NUM}$ for all $t \in \mathcal{R}$ and all $[\varphi, \Phi]$. On the other hand, there are complexity measures $[\varphi, \Phi]$ and $t \in \mathcal{R}$ such that $\mathcal{C}_t^{[\varphi,\Phi]} \notin \text{NUM}!$.

Theorem 15.16 *Let $[\varphi, \Phi]$ be any complexity measure. Then, for every function $t \in \mathcal{R}$ we have $\mathcal{C}_t^{[\varphi,\Phi]} \in \text{NUM}$.*

Proof. We start our proof by asking under which conditions a function f belongs to $\mathcal{C}_t^{[\varphi,\Phi]}$. By the definition of a complexity class, this is the case if and only if there is a program k such that $\varphi_k = f$ and Φ_k satisfies

$$\forall^\infty n [\Phi_k(n) \leq t(n)]. \tag{15.4}$$

Condition (15.4) means that there exists a $\tau \in \mathbb{N}$ and an $n_0 \in \mathbb{N}$ such that $\Phi_k(n) \leq \tau$ for all $n < n_0$ and $\Phi_k(n) \leq t(n)$ for all $n \geq n_0$.

Therefore, we choose an effective enumeration of all triples of natural numbers. For the sake of presentation, let $c_1, c_2, c_3 \in \mathcal{R}$ such that

$$\mathbb{N} \times \mathbb{N} \times \mathbb{N} = \{[c_1(i), c_2(i), c_3(i)] \mid i \in \mathbb{N}\} \, .$$

Next, we define the desired function $g \in \mathcal{R}^2$. For all $i, n \in \mathbb{N}$ let

$$g(i,n) = \begin{cases} \varphi_{c_1(i)}(n), & \text{if } [n < c_2(i) \wedge \Phi_{c_1(i)}(n) \leq c_3(i)] \\ & \vee [n \geq c_2(i) \wedge \Phi_{c_1(i)}(n) \leq t(n)] \, ; \\ 0, & \text{otherwise} \, . \end{cases}$$

It remains to show that $\mathcal{C}_t^{[\varphi, \Phi]} \subseteq \{\lambda n.g(i,n) \mid i \in \mathbb{N}\}$. Let $f \in \mathcal{C}_t^{[\varphi, \Phi]}$; then there exists a triple $[k, n_0, \tau]$ such that

(1) $\varphi_k = f$,

(2) $\forall n < n_0 [\Phi_k(n) \leq \tau]$,

(3) $\forall n \geq n_0 [\Phi_k(n) \leq t(n)]$.

Hence, there must exist an $i \in \mathbb{N}$ such that $[c_1(i), c_2(i), c_3(i)] = [k, n_0, \tau]$. Now, by construction we can conclude that $\lambda n.g(i,n) = \varphi_k = f$, and thus, by (1) through (3), the theorem follows. ∎

The proof above points to the difficulty to show $\mathcal{C}_t^{[\varphi, \Phi]} \in \text{NUM}!$, i.e., we have set $g(i,n) = 0$ if the stated condition is not satisfied. Since we have to ensure $g \in \mathcal{R}^2$, we must define $g(i,n)$ somehow for all those i, n for which the stated condition is not fulfilled. So we run into the danger to enumerate functions that are *not* elements of $\mathcal{C}_t^{[\varphi, \Phi]}$. But if we have a bit prior knowledge about $\mathcal{C}_t^{[\varphi, \Phi]}$, then we can show $\mathcal{C}_t^{[\varphi, \Phi]} \in \text{NUM}!$. We set

$$U_0 = \{f \mid f \in \mathcal{R}, \, f(n) = 0 \text{ for all but finitely many } n\} \, ,$$

i.e., U_0 is the class of functions of *finite support*. For any $f \in \mathcal{R}$ we define

15.4 Complexity Classes

$$U_f = \{\hat{f} \mid \hat{f} \in \mathcal{R},\ \hat{f}(n) = f(n) \text{ for all but finitely many } n\},$$

i.e., U_f is the class of all *finite variations of function* f.

Theorem 15.17 *Let $[\varphi, \Phi]$ be any complexity measure and let $t \in \mathcal{R}$. Then we have:*

(1) *If $U_0 \subseteq \mathcal{C}_t^{[\varphi,\Phi]}$ then $\mathcal{C}_t^{[\varphi,\Phi]} \in \text{NUM!}$.*

(2) *If $U_f \subseteq \mathcal{C}_t^{[\varphi,\Phi]}$ for some $f \in \mathcal{R}$ then $\mathcal{C}_t^{[\varphi,\Phi]} \in \text{NUM!}$.*

Proof. The proof is conceptually similar to the proof of Theorem 15.16. Only one important modification is necessary. First, we show Assertion (1). Let $c_1, c_2, c_3 \in \mathcal{R}$ be as in the proof of Theorem 15.16. Now, we define the desired function $g \in \mathcal{R}^2$ as follows. For all $i, n \in \mathbb{N}$ let

$$g(i,n) = \begin{cases} \varphi_{c_1(i)}(n), & \text{if } [\forall x[x < c_2(i) \implies \Phi_{c_1(i)}(x) \leq c_3(i)] \land \\ & \forall x[c_2(i) \leq x \leq n \implies \Phi_{c_1(i)}(x) \leq t(x)]]\ ; \\ 0, & \text{otherwise}\ . \end{cases}$$

One easily verifies that for all $i \in \mathbb{N}$ the condition $\lambda n.g(i,n) = \varphi_{c_1(i)}$ or $\forall^\infty n\, g(i,n) = 0$ is satisfied. Consequently, now Assertion (1) follows as in the proof of Theorem 15.16. We omit the details.

Assertion (2) is proved *mutatis mutandis*, i.e., we define for all $i, n \in \mathbb{N}$

$$g(i,n) = \begin{cases} \varphi_{c_1(i)}(n), & \text{if } [\forall x[x < c_2(i) \implies \Phi_{c_1(i)}(x) \leq c_3(i)] \land \\ & \forall x[c_2(i) \leq x \leq n \implies \Phi_{c_1(i)}(x) \leq t(x)]]\ ; \\ f(n), & \text{otherwise}\ . \end{cases}$$

We omit the details. ∎

Next, we can show the following general result.

Theorem 15.18 *Let $[\varphi, \Phi]$ be any complexity measure. Then there exists a function $t \in \mathcal{R}$ such that, for all functions $\tilde{t}$ satisfying $\tilde{t}(n) \geq t(n)$ for all but finitely many n, we have $\mathcal{C}_{\tilde{t}}^{[\varphi,\Phi]} \in \text{NUM!}$.*

Proof. The proof is done via the following claim.

Claim. *Let $[\varphi, \Phi]$ be any complexity measure. Then there exists a function $\hat{t} \in \mathcal{R}$ such that $U_0 \subseteq \mathcal{C}_{\hat{t}}^{[\varphi, \Phi]}$.*

First, note that $U_0 \in \text{NUM}$, since all finite tuples of natural numbers are recursively enumerable. Thus, we can apply Theorem 15.15. Consequently, there is a function $\hat{t} \in \mathcal{R}$ such that $U_0 \subseteq \mathcal{C}_{\hat{t}}^{[\varphi, \Phi]}$. This proves the claim.

Now, we set $t = \hat{t}$ and get $U_0 \subseteq \mathcal{C}_{t}^{[\varphi, \Phi]}$. Assume any function $\tilde{t}$ satisfying $\tilde{t}(n) \geq t(n)$ for all but finitely many n. By the definition of complexity classes, we get $\mathcal{C}_{t}^{[\varphi, \Phi]} \subseteq \mathcal{C}_{\tilde{t}}^{[\varphi, \Phi]}$, and therefore we have $U_0 \subseteq \mathcal{C}_{\tilde{t}}^{[\varphi, \Phi]}$, too. Thus, by Theorem 15.17 we arrive at $\mathcal{C}_{\tilde{t}}^{[\varphi, \Phi]} \in \text{NUM}!$. ∎

The latter theorem allows for a nice corollary. Consider all 3-tape Turing machines (with input-tape, work-tape, and output tape) and let $\tilde{\varphi}$ be the canonical Gödel numbering of all these 3-tape Turing machines. Moreover, we let Φ_i be the number of cells used by φ_i on its work-tape. Thus, $[\tilde{\varphi}, \tilde{\Phi}]$ is a complexity measure.

Corollary 15.5 *Let $[\tilde{\varphi}, \tilde{\Phi}]$ be the complexity measure defined above. Then $\mathcal{C}_{t}^{[\tilde{\varphi}, \tilde{\Phi}]} \in \text{NUM}!$ for all $t \in \mathcal{R}$.*

Proof. The corollary is a direct consequence of Theorem 15.18, since $U_0 \subseteq \mathcal{C}_{t}^{[\tilde{\varphi}, \tilde{\Phi}]}$ for all $t \in \mathcal{R}$, i.e, in particular for $t(n) = 0$ for all n. ∎

Next we show that the theorems concerning the recursive enumerability of complexity classes cannot be improved. That is, we prove that there are complexity measures $[\varphi, \Phi]$ and functions $t \in \mathcal{R}$ such that $\mathcal{C}_{t}^{[\varphi, \Phi]} \notin \text{NUM}!$. For the sake of presentation, it will be convenient to identify a function with the sequence of its values. For example, we shall use 0^∞ to denote the constant zero function.

Theorem 15.19 *There is a complexity measure $[\varphi, \Phi]$ and a function $t \in \mathcal{R}$ such that $\mathcal{C}_{t}^{[\varphi, \Phi]} \notin \text{NUM}!$.*

15.4 Complexity Classes

Proof. Since the theorem can only hold for small functions $t \in \mathcal{R}$, we choose the smallest possible one, i.e., $t = 0^\infty$. Now, let $[\varphi, \Phi']$ be any complexity measure. Furthermore, let $h \in \mathcal{R}$ be a function such that

$$\varphi_{h(i)}(n) = i \quad \text{for all } i \in \mathbb{N} \text{ and all } n \in \mathbb{N}.$$

That is, $\varphi_{h(i)} = i^\infty$. Without loss of generality, we can assume h to be strictly monotonic. Thus, $range(h)$ is decidable (cf. Problem 14.3).

Now, the proof idea is easily explained. Let K be the halting set. Recall that K is recursively enumerable but $\overline{K}$ is not (cf. Corollaries 14.6 and 14.7). We construct a complexity measure $[\varphi, \Phi]$ such that

$$C_{0^\infty}^{[\varphi,\Phi]} = \{i^\infty \mid i \in \overline{K}\}. \tag{15.5}$$

This is all we need. For seeing this, assume we have already shown Property (15.5) to hold. Suppose to the contrary that $C_{0^\infty}^{[\varphi,\Phi]} \in \text{NUM!}$. Then there must be a function $g \in \mathcal{R}^2$ such that

$$C_{0^\infty}^{[\varphi,\Phi]} = \{\lambda x.g(j,x) \mid j \in \mathbb{N}\}.$$

The latter property directly implies that $\{g(j,0) \mid j \in \mathbb{N}\} = \overline{K}$, i.e., $\overline{K}$ would be recursively enumerable, a contradiction.

So it remains to show Property (15.5). For all $j, n \in \mathbb{N}$ we set

$$\Phi_j(n) = \begin{cases} 0, & \text{if } j = h(i) \text{ and } \neg[\Phi'_i(i) \leq n]; \\ 1 + \Phi'_j(n), & \text{if } j = h(i) \text{ and } \Phi'_i(i) \leq n; \\ 1 + \Phi'_j(n), & \text{if } j \notin range(h). \end{cases}$$

In the first and second case we have $j \in range(h)$. Since h is strictly monotonic, there can be at most one such i satisfying $j = h(i)$. In the third case, we assume $\Phi_j(n) \uparrow$ if and only if $\Phi'_j(n) \uparrow$.

It is easy to see that $[\varphi, \Phi]$ is a complexity measure. Since we did not change φ, our construction implies $\varphi \in \text{Göd}$, so Condition (1) of Definition 15.4 holds.

Condition (2) of Definition 15.4 is also satisfied. For deciding whether or not "$\Phi_j(n) = y$," we have to check if $j \in range(h)$. This is decidable. If $j \notin range(h)$ and $y = 0$ then the answer is "no." If $j \notin range(h)$ and $y > 0$, we output $M'(j, n, y-1)$, where M' is the predicate for $[\varphi, \Phi']$. Moreover, if $j \in range(h)$, then let $j = h(i)$. We then first check if $M'(i, i, k) = 0$ for all $k = 0, \ldots, n$. If this true and $y = 0$, we output 1. If $M'(i, i, k) = 0$ for all $k = 0, \ldots, n$ and $y > 0$, we output 0. Otherwise, there is a $k \leq n$ such that $M'(i, i, k) = 1$. Again, if $y = 0$, we output 0, and if $y > 0$ we output $M'(j, n, y-1)$.

Finally, it is easy to see that for all $j \in \mathbb{N}$ our construction directly yields $\Phi_j(n) > 0$ for all but finitely many n or $\Phi_j(n) = 0$ for all $n \in \mathbb{N}$. But the latter case happens if and only if $\exists i [j = h(i) \wedge \varphi_i(i) \uparrow]$, i.e., $\mathcal{C}_{0\infty}^{[\varphi, \Phi]} = \{i^\infty \mid i \in \overline{K}\}$. This proves the theorem. ∎

15.4.2 An Undecidability Result

As we have seen, it can happen that $\mathcal{C}_t^{[\varphi, \Phi]} \notin$ NUM!. Thus, it is only natural to ask if one can decide, for any given complexity measure $[\varphi, \Phi]$ and any given function $t \in \mathcal{R}$ whether or not $\mathcal{C}_t^{[\varphi, \Phi]} \in$ NUM!. Our next theorem shows that there is such a decision procedure if and only if it is not necessary, i.e., if and only if $\mathcal{C}_t^{[\varphi, \Phi]} \in$ NUM! for all $t \in \mathcal{R}$. Thus, conceptually, the following theorem should remind us to Rice's theorem.

Theorem 15.20 *Let $[\varphi, \Phi]$ be any complexity measure. Then we have:*

$$\exists \psi \in \mathcal{P} \, \forall j [\varphi_j \in \mathcal{R} \implies \psi(j) \text{ defined} \wedge \left(\mathcal{C}_{\varphi_j}^{[\varphi, \Phi]} \in \text{NUM!} \implies \psi(j) = 1 \right)$$
$$\wedge \left(\mathcal{C}_{\varphi_j}^{[\varphi, \Phi]} \notin \text{NUM!} \implies \psi(j) = 0 \right) \iff \forall t \in \mathcal{R} \left[\mathcal{C}_t^{[\varphi, \Phi]} \in \text{NUM!} \right] .$$

Proof. The sufficiency is obvious, since we can set $\psi = 1^\infty$.

Necessity. We set $\mathcal{R}^+ = \{t \mid t \in \mathcal{R} \wedge \mathcal{C}_t^{[\varphi, \Phi]} \in \text{NUM!}\}$ as well as $\mathcal{R}^- = \mathcal{R} \setminus \mathcal{R}^+$. As we have seen, $\mathcal{R}^+ \neq \emptyset$ (cf. Theorem 15.18). Moreover, if

$\mathcal{R}^- = \emptyset$, then we are already done. Thus, suppose $\mathcal{R}^- \neq \emptyset$. We continue by showing that such a desired ψ cannot exist. For this purpose, let $t^+ \in \mathcal{R}^+$ and $t^- \in \mathcal{R}^-$. Furthermore, for any $f \in \mathcal{R}$ we again set

$$U_f = \{\hat{f} \mid \hat{f} \in \mathcal{R}, \ \hat{f}(n) = f(n) \text{ for all but finitely many } n\} \ .$$

Now, suppose to the contrary that there is a $\psi \in \mathcal{P}$ such that

$$\forall j [\varphi_j \in \mathcal{R} \implies \psi(j) \text{ defined} \wedge \left(\mathcal{C}^{[\varphi, \Phi]}_{\varphi_j} \in \text{NUM!} \implies \psi(j) = 1 \right)$$
$$\wedge \left(\mathcal{C}^{[\varphi, \Phi]}_{\varphi_j} \notin \text{NUM!} \implies \psi(j) = 0 \right) \ .$$

Then, this function ψ should in particular satisfy the following

$$\forall j [\varphi_j \in U_{t^+} \implies \psi(j) = 1] \wedge \forall j [\varphi_j \in U_{t^-} \implies \psi(j) = 0] \ .$$

Next, let k be arbitrarily fixed such the $\varphi_k = \psi$. Furthermore, let $f \in \mathcal{R}$ be chosen such that for all $j, n \in \mathbb{N}$

$$\varphi_{f(j)}(n) = \begin{cases} t^+(n), & \text{if } \Phi_k(j) \leq n \wedge \psi(j) = 0 \ ; \\ t^-(n), & \text{if } \Phi_k(j) \leq n \wedge \psi(j) = 1 \ ; \\ 0, & \text{otherwise} \ . \end{cases}$$

By the choice of k, if $\Phi_k(j) \leq n$, then we can compute $\varphi_k(j) = \psi(j)$. Thus, $\varphi_{f(j)} \in \mathcal{R}$ for all $j \in \mathbb{N}$. By Theorem 15.5, there is an $a \in \mathbb{N}$ such that $\varphi_{f(a)} = \varphi_a$. Hence, $\varphi_a \in \mathcal{R}$. By assumption, we can conclude that $\psi(a) \downarrow$. We distinguish the following cases.

Case 1. $\psi(a) = 0$.

By construction, $\varphi_{f(a)} \in U_{t^+}$. Since $\varphi_{f(a)} = \varphi_a$, we get $\varphi_a \in U_{t^+}$. But this implies $\mathcal{C}^{[\varphi, \Phi]}_{\varphi_a} \in \text{NUM!}$, and consequently $\psi(a) = 1$, a contradiction. Thus, Case 1 cannot happen.

Case 2. $\psi(a) = 1$.

By construction, $\varphi_{f(a)} \in U_{t^-}$ and thus $\varphi_a \in U_{t^-}$. By the choice of t^-, we get $\mathcal{C}_{\varphi_a}^{[\varphi,\Phi]} \notin \text{NUM!}$, and therefore $\psi(a) = 0$, a contradiction. So Case 2 cannot happen either and the desired function ψ cannot exist. ∎

Further results concerning recursive properties of abstract complexity classes can be found in Landweber and Robertson [22] as well as in Mc-Creight and Meyer [25] and the references therein.

15.4.3 The Gap-Theorem

Next, we turn our attention to a different problem. We are going to ask how much resources must be added to some given resources in order to make more functions computable than before. That is, given any $t \in \mathcal{R}$, we are interested in learning how to choose $\hat{t}$ such that $\mathcal{C}_t^{[\varphi,\Phi]} \subset \mathcal{C}_{\hat{t}}^{[\varphi,\Phi]}$.

Before we are trying to answer this question in more depth, we confine ourselves that there is always such a function $\hat{t}$.

Lemma 15.1 *Let $[\varphi, \Phi]$ be any complexity measure. For every function $t \in \mathcal{R}$ there is a function $\hat{t}$ such that $\mathcal{C}_t^{[\varphi,\Phi]} \subset \mathcal{C}_{\hat{t}}^{[\varphi,\Phi]}$.*

Proof. Let $t \in \mathcal{R}$ be arbitrarily fixed. By Theorem 15.11, there exists a function $f \in \mathcal{R}_{0,1}$ such that for all programs i with $\varphi_i = f$ we have $\Phi_i(n) > t(n)$ for all but finitely many $n \in \mathbb{N}$. Therefore, we get $f \notin \mathcal{C}_t^{[\varphi,\Phi]}$. Let i be any program for f. We set $\hat{t}(n) = \max\{t(n), \Phi_i(n)\}$ for all $n \in \mathbb{N}$. Consequently, $f \in \mathcal{C}_{\hat{t}}^{[\varphi,\Phi]}$ and $\mathcal{C}_t^{[\varphi,\Phi]} \subset \mathcal{C}_{\hat{t}}^{[\varphi,\Phi]}$. ∎

The latter lemma directly implies the following corollary.

Corollary 15.6 *Let $[\varphi, \Phi]$ be any complexity measure. There is no function $t \in \mathcal{R}$ such that $\mathcal{C}_t^{[\varphi,\Phi]} = \mathcal{R}$.*

So, the more interesting question is whether or not we can choose $\hat{t}$ in dependence of t effectively in order to obtain $\mathcal{C}_t^{[\varphi,\Phi]} \subset \mathcal{C}_{\hat{t}}^{[\varphi,\Phi]}$.

Let us try the following approach. For the sake of presentation, for any

$h \in \mathcal{R}^2$ we shall write $h \circ t$ to denote the function $(h \circ t)(n) = h(n, t(n))$ for all $n \in \mathbb{N}$. Now, we can imagine that $h \in \mathcal{R}^2$ is a very rapidly growing function. Thus, it is quite natural to ask whether or not we may expect

$$\mathcal{C}_t^{[\varphi,\Phi]} \subset \mathcal{C}_{h \circ t}^{[\varphi,\Phi]}$$

for all $t \in \mathcal{R}$. Surprisingly, the answer is *no* as the following theorem shows. Since this theorem establishes a "gap" in which nothing more can be computed than before, it is called Gap-theorem. The Gap-theorem was discovered by Trakhtenbrot [37] and later independently by Borodin [3]. The proof given below, however, follows Young [39].

Theorem 15.21 (Gap-Theorem) *Let $[\varphi, \Phi]$ be any complexity measure. Then we have*

$$\forall h \in \mathcal{R}^2 \left[\forall n \forall y [h(n, y) \geq y \implies \exists t \in \mathcal{R} \left[\mathcal{C}_t^{[\varphi,\Phi]} = \mathcal{C}_{h \circ t}^{[\varphi,\Phi]} \right] \right] .$$

Proof. We shall even show a somehow stronger result, i.e., that t can be made arbitrarily large. Let $a \in \mathcal{R}$ be any function. We are going to construct t such that

(1) $t(n) \geq a(n)$ for all but finitely many n,
(2) for all $n > j$, if $\Phi_j(n) > t(n)$ then $\Phi_j(n) > h(n, t(n))$.

To define t we set

$$t_{n+1} = a(n) \text{ and for } 0 < i \leq n+1, \text{ let } t_{i-1} = h(n, t_i) + 1 .$$

Thus, we directly get from $h(n, y) \geq y$ that

$$t_{n+1} < t_n < t_{n-1} < \cdots < t_1 < t_0 .$$

So, we have $n + 2$ many points. We consider all the $n + 1$ many intervals $[t_i, t_{i-1})$, $i = n+1, \ldots, 1$. Moreover, consider the n many points $\Phi_j(n)$

for $j = 0, \ldots, n-1$. Since we have more intervals than points, we can effectively find at least one interval, say $[t_{i_0}, t_{i_0-1})$ that does not contain any of these points, i.e., for no $j < n$ do we have

$$t_{i_0} \leq \Phi_j(n) \leq h(n, t_{i_0}) < t_{i_0-1} .$$

Therefore, we set $t(n) = t_{i_0}$. By construction $t(n) = t_{i_0} \geq t_{n+1} = a(n)$. For $n > j$ and $\Phi_j(n) \geq t(n) = t_{i_0}$ we obtain from $\Phi_j(n) \notin [t_{i_0}, t_{i_0-1})$ that

$$\Phi_j(n) \geq t_{i_0-1} = h(n, t_{i_0}) + 1 > h(n, t(n)) .$$

Since $h(n, y) \geq y$, the condition $\Phi_j(n) \leq t(n)$ implies $\Phi_j(n) \leq h(n, t(n))$. Consequently, $C_t^{[\varphi, \Phi]} = C_{h \circ t}^{[\varphi, \Phi]}$. ∎

So, we have just seen that there are indeed gaps in the complexity hierarchy which do not contain any new function from $\mathcal{R}$. These gaps are described by the functions t and $h \circ t$ from the Gap-Theorem.

15.5 Problem Set 15

Problem 15.1 Let $[\varphi, \Phi]$ be any complexity measure and let $U_{[\varphi, \Phi]} = \{\Phi_i \mid \varphi_i \in \mathcal{R}\}$. Prove or disprove that $U_{[\varphi, \Phi]} \in \text{NUM}$.

Problem 15.2 Let $INF = \{f \mid f \in \mathcal{P}, \ dom(f) \text{ is infinite}\}$. Prove or disprove that there is a numbering $\psi \in \mathcal{P}^2$ such that $INF = \{\psi_i \mid i \in \mathbb{N}\}$.

Problem 15.3 Let $\varphi \in \text{Göd}$, let e be the nowhere defined function and let $A = \{i \mid i \in \mathbb{N}, \ dom(\varphi_i) = dom(e)\}$. Prove or disprove A to be recursively enumerable.

Appendix

A.1 Greek Alphabet （ギリシャ文字）

Greek Alphabet

A	α	alpha	（アルファ）	N	ν	nu	（ニュー）
B	β	beta	（ベータ）	Ξ	ξ	xi	（クスィー）
Γ	γ	gamma	（ガンマ）	O	o	omicron	（オミクロン）
Δ	δ	delta	（デルタ）	Π	π	pi	（パイ）
E	ϵ	epsilon	（エプシロン）	P	ρ	rho	（ロー）
Z	ζ	zeta	（ジータ）	Σ	σ	sigma	（シグマ）
H	η	eta	（イータ）	T	τ	tau	（タウ）
Θ	θ	theta	（シータ）	Υ	υ	upsilon	（ウプシロン）
I	ι	iota	（イオタ）	Φ	φ	phi	（ファイ）
K	κ	kappa	（カッパ）	X	χ	chi	（キー、カイ）
Λ	λ	lambda	（ラムダ）	Ψ	ψ	psi	（プサイ）
M	μ	mu	（ミュー）	Ω	ω	omega	（オメガ）

Bibliography

[1] Wilhelm Ackermann. Zum Hilbertschen Aufbau der reellen Zahlen. *Mathematische Annalen*, 99:118–133, 1928.

[2] Manuel Blum. A machine independent theory of the complexity of recursive functions. *Journal of the ACM*, 14(2):322–336, 1967.

[3] A. Borodin. Computational complexity and the existence of complexity gaps. *Journal of the ACM*, 19(1):158–174, 1972.

[4] N. Chomsky and M. P. Schützenberger. The algebraic theory of context-free languages. In P. Braffort and D. Hirschberg, editors, *Computer Programming and Formal Languages*, pages 118–161. North-Holland Publishing Company, Amsterdam, 1963.

[5] Alonzo Church. An unresolvable problem of elementary number theory. *American Journal of Mathematics*, 58:345–365, 1936.

[6] Maxime Crochemore and Wojciech Rytter. *Jewels of Stringology, Text Algorithms*. World Scientific, New Jersey, London, Singapore Hong Kong, 2003.

[7] Calvin C. Elgot and Abraham Robinson. Random access stored program machines, an approach to programming languages. *Journal of the ACM*, 11(4):365–399, 1964.

[8] Richard M. Friedberg. Three theorems on recursive enumeration. *The Journal of Symbolic Logic*, 23(3):309–316, 1958.

[9] Seymour Ginsburg. *The Mathematical Theory of Context-Free Languages*. McGraw-Hill, Inc, New York, NY, USA, 1966.

[10] Kurt Gödel. Über formal unentscheidbare Sätze der Principia Mathematica und verwandter Systeme. *Monatshefte Mathematik Physik*, 38:173–198, 1931.

[11] Sheila A. Greibach. A new normal-form theorem for context-free phrase structure grammars. *Journal of the ACM*, 12(1):42–52, 1965.

[12] Michael A. Harrison. *Introduction to Formal Language Theory*. Addison–Wesley Publishing Company, Reading Massachusetts, 1978.

Bibliography   209

[13] Juris Hartmanis. Computational complexity of random access stored program machines. *Mathematical Systems Theory*, 5(3):232–245, 1971.

[14] Hans Hermes. *Aufzählbarkeit, Entscheidbarkeit, Berechenbarkeit: Einführung in die Theorie der Rekursiven Funktionen*. Springer-Verlag, Berlin, Heidelberg, New York, second edition, 1971.

[15] John E. Hopcroft, Rajeev Motwani, and Jeffrey D. Ullman. *Introduction to Automata Theory, Languages and Computation*. Addison–Wesley Publishing Company, Reading Massachusetts, second edition, 2001.

[16] John E. Hopcroft and Jeffrey D. Ullman. *Introduction to Automata Theory, Languages and Computation*. Addison–Wesley Publishing Company, Reading Massachusetts, 1979.

[17] Neil Immerman. Nondeterministic space is closed under complementation. *SIAM Journal of Computing*, 17(5):935–938, 1988.

[18] S. C. Kleene. On notation for ordinal numbers. *The Journal of Symbolic Logic*, 3(4):150–155, 1938.

[19] S. C. Kleene. Recursive predicates and quantifiers. *Transactions of the American Mathematical Society*, 53(1):41–73, 1943.

[20] Martin Kummer. An easy priority-free proof of a theorem of Friedberg. *Theoretical Computer Science*, 74(2):249–251, 1990.

[21] S.-Y. Kuroda. Classes of languages and linear-bounded automata. *Information and Control*, 7(2):207–223, 1964.

[22] L. H. Landweber and E. L. Robertson. Recursive properties of abstract complexity classes. *Journal of the ACM*, 19(2):296–308, 1972.

[23] A. A. Markov. *Theory of Algorithms*. Academy of Sciences of the USSR, Tr. Mat. Inst. Steklov, No. 42, Moscow, 1954.

[24] Yuri V. Matiyasevich. *Hilbert's Tenth Problem*. MIT Press, Cambridge, Massachusetts, USA, 1993.

[25] E. M. McCreight and A. R. Meyer. Classes of computable functions defined by bounds on computation: Preliminary report. In *Proceedings of the first annual ACM symposium on Theory of computing, Marina del Rey, California, United States*, pages 79–88. ACM Press, 1969.

[26] Anil Nerode. Linear automaton transformations. *Proceedings of the American Mathematical Society*, 9:541–544, 1958.

[27] Rózsa Péter. *Recursive Functions in Computer Science*. Akadémiai Kiadó,

Budapest, Hungary, 1981.

[28] Emil L. Post. Formal reductions of the general combinatorial decision problem. *American Journal of Mathematics*, 65:197–215, 1943.

[29] Emil L. Post. A variant of a recursively unsolvable problem. *Bulletin of the American Mathematical Society*, 52:264–268, 1946.

[30] M. O. Rabin and D. Scott. Finite automata and their decision problems. *IBM Journal of Research and Development*, 3:114–125, 1959.

[31] H. G. Rice. Classes of recursively enumerable sets and their decision problems. *Transactions of the American Mathematical Society*, 74(2):358–366, 1953.

[32] Hartley Rogers, Jr. Gödel numberings of partial recursive functions. *The Journal of Symbolic Logic*, 23(3):331–341, 1958.

[33] Hartley Rogers, Jr. *Theory of Recursive Functions and Effective Computability*. MIT Press, Cambridge, Massachusetts, USA, 1987.

[34] John C. Shepherdson and Howard E. Sturgis. Computability of recursive functions. *Journal of the ACM*, 10(2):217–255, 1963.

[35] Raymond M. Smullyan. *Theory of formal systems*, volume 47 of *Annals of Mathematics Studies*. Princeton University Press, Princeton, New Jersey, USA, 1961.

[36] Róbert Szelepcsényi. The method of forced enumeration for nondeterministic automata. *Acta Informatica*, 26(3):279–284, 1988.

[37] B. A. Trakhtenbrot. Turing computations with logarithmic delay. *Algebra i Logika*, 3(4):33–48, 1964.

[38] A. M. Turing. On computable numbers with an application to the Entscheidungsproblem. *Proceedings of the London Mathematical Society*, 42(2):230–265, 1936.

[39] Paul Young. Easy constructions in complexity theory: Gap and speed-up theorems. *Proc. of the American Mathematical Society*, 37(2):555–563, 1973.

Subject Index

【A】

acceptance
  via empty stack 96
  via final state 96
acceptor (受理器) 2
Ackermann, Wilhelm   134
action (アクション) .. 61
algorithm 46
  Boyer-Moore 46
  CNF 71
  GNF 107
  intuitive notion 120
  Knuth-Morris-Pratt   46
  separated grammar   69
algorithmically unsolvable
  (計算論的に解決不能)   155
alphabet
  (アルファベット) .. 7
  tape-symbols
    (テープ記号)   138
Al-Hwarizmi 119
ambiguity
  (曖昧さ) 64
arithmetic difference
  (数論的減算) 128
Ars magna 119
Art inveniendi 119
associative (結合的)   7, 68
at most countably infinite
  (高々可算無限) 4
automaton (オートマトン)
  see finite automaton

【B】

Backus normal form ......
  see BNF
Backus, John 59
Backus-Naur Form ....
  see BNF
Bar-Hillel, Yehoshua .. 72
binary relation
  (二項関係) 5
  antisymmetric
    (反対称的) 5
  equality (恒等) 6
  reflexive (反射的) ... 5
  symmetric (対称的)   5
  transitive (推移的)   5
Blum, Manuel 188
BNF 59
Borodin, Allan 205

【C】

cancellation
  (無効化) 191
Cantor's
  pairing function ... 133
  theorem 4, 121
Cantor, Georg 133
cardinality (要素数) .. 2
$\mathcal{CF}$ 48
$\mathcal{CF}$ closed under
  homomorphisms 82
  intersection w. $\mathcal{REG}$ 54,
  104, (263)
  Kleene closure 49

product 49
substitution 78
transposition 52
union 49
$\mathcal{CF}$ not closed under
  complement 53
  intersection 53
characteristic function
  (特性関数) .. 130, 156
characterization
  of $\mathcal{CF}$ 113
  of $\mathcal{CS}$ 115
  of $\mathcal{REG}$ 25
  partial rec. funct. .. 140
  recursive set 176
  Turing comp. funct.   140
character class
  (文字クラス) 41
Chomsky hierarchy
  (チョムスキー階層)   117
Chomsky normal form
  (チョムスキー標準形)
  see CNF
Chomsky, Noam .. 48, 83
Chomsky-Schützenberger
  theorem
  (チョムスキー-シュツェン
  ベルガーの定理) 83
Church
  λ-calculus 145
Church's thesis
  (チャーチの提唱)   145
Church, Alonzo 145
closure properties ....

see $\mathcal{CF}, \mathcal{CS}, \mathcal{REG}$
closure（閉包）.......... *6*
CNF *70, 78, 83, 106*
   definition *70*
   properties *70*
compiler（コンパイラ）*182*
complement（補集合）*53*
complexity class
   （計算量クラス）.. *195*
   in NUM! *199*
   in NUM *197*
complexity measure
   （計算量尺度）..... *188*
   carbon *189*
   definition *188*
   ink *189*
   recursive relatedness *190*
   reversal *189*
   space *189*
   time *189*
composition（合成）*122*
   of binary relations ... *5*
computability
   （計算可能性）..... *113*
computation
   accepting *23*
concatenation（連接）
   definition *7*
   properties of *8*
configuration ....
   see TM
context-free
   （文脈自由）......... *48*
context-sensitive
   （文脈依存）......... *114*
contradiction（矛盾）.. *5*
countably infinite
   （可算無限）.... *4, 121*
$\mathcal{CS}$ closed under
   intersection *114*
   Kleene closure *114*
   product *114*
   transposition *114, (265)*

union *114*

【D】

decidability *39*
   context-free lang.
     $\lambda \in L(\mathcal{G})$ ... *(254)*
     $L(\mathcal{G})$ finite *180*
     $L(\mathcal{G})$ infinite . *180*
     $L(\mathcal{G}) = \emptyset$ *56*
     $s \in L(\mathcal{G})$ *180*
   context-sensitive lang.
     $s \in L(\mathcal{G})$ *116*
   regular languages ... *39*
     equality *175*
     subset *175*
     $L(\mathcal{G})$ finite *39*
     $L(\mathcal{G})$ infinite ... *39*
     $L(\mathcal{G}) = \emptyset$ *174*
     $s \in L(\mathcal{G})$ *27*
decidable（決定可能）*156*
decision procedure
   （決定手続き）..... *119*
Dedekind
   justification theorem *123*
Dedekind, Richard .. *123*
definition
   class $\mathcal{P}$ *126*
   class *Prim* *126*
   parse tree *61*
derivation（導出）..... *15*
   leftmost *67*
   leftmost（最左導出）*67*
   rightmost（最右導出）*67*
DFA *22, 29, 32, 46, (232)*
$L(\mathcal{A}) \cap L(\tilde{\mathcal{A}})$ .... *(256)*
minimization ..... *(246)*
diagonalization
   （対角化）..... *154, 191*
directly generate
   （直接生成する）.... *15*
domain（定義域）... *140*
Dyck language
   （ダイク言語）.. *82, 83*

Dyck, Walter von *82*

【E】

empty string
   （空文字列）.......... *7*
enumeration procedure
   （列挙手続き）..... *119*
equivalence class
   （同値類）............. *5*
equivalence relation
   （同値関係）.......... *5*
equivalence（等性）
   of grammars *19*
equivalent（等価）..... *19*
Euclid's Elements
   （ユークリッド原論）*118*

【F】

Fibonacci seq. ...... *(258)*
final state(s)
   （最終状態）
   DFA *22*
   NFA *22*
finite automaton
   （有限オートマトン）*2, 22, 25*
deterministic
   （決定性）...... *22*
equivalence
   DFA NFA *25*
nondeterministic
   （非決定性）.... *22*
finite extension
   （有限拡大）....... *191*
finite rank
   （有限の階数）....... *31*
fixed point theorem
   （不動点定理）..... *184*
flex *43, 44*
free monoid
   （自由モノイド）..... *7*
function（関数）.. *3, 121*
   Ackermann-Péter ... *134*

arithmetic difference
 (数論的減算) 128
basic functions 121
bijective (全単射) ... 3
binary addition 126
binary multiplication 128
by case distinction 130
characteristic (特性)
    of a set 156
    of predicate ... 130
computed by TM 139
constant 0
 (定数 0 関数) 122
general addition ... 129
general multiplication 129
general recursive
 (一般帰納的関数) 133
identity
 (恒等関数) 126
injective (単射) 3
noncomputable 121
nowhere defined ... 134
pairing 131, 146
partial characteristic
 (部分特性関数) ... 176
partial recursive
 (部分帰納的関数) 121,
126
predecessor
 (前者関数) 122
primitive recursive
 (原始帰納的関数) 126
signum
 (符号関数) 128
strong monotonic 181
successor
 (後者関数) 122
surjective (全射) 3
h-honest (h-正当) 193

【G】

gap-theorem 205
general recursive ....

see function (関数)
generation (生成) 15
direct 15
GNF 105, 111
    example 108
grammar (文法) ... 2, 14
    ambiguous 64, 68
    CNF 70
    context-free
    (文脈自由文法) 48
    context-sensitive
    (文脈依存文法) ... 114
    equivalent 19
    GNF 105, 106
    λ-free (λ-自由) ... 57,
    (252)
    length-increasing
    (非縮小文法) 115
    reduced (既約) 54, 55
    regular (正規文法) 16
    $L(\mathcal{G}) \cap L(\tilde{\mathcal{G}})$ (255)
    separated
    (分離的文法) 69
    type-0 ...... 15, 82, 178
    unambiguous 68
Greibach normal form
 (グライバッハ標準形)
 see GNF
Greibach, Sheila A. . 105
grep 41, 45
Gödel numbering ....
 see numbering
Gödel (ゲーデル) ... 183
Gödel, Kurt 121, 183

【H】

halting problem (停止問題)
    general (一般) 154
    undecidable ... 156
    undecidable 154
halting set (停止集合) 157,
201
    not recursive 177

recursively enumerable
177
head (ヘッド) 23
Hilbert's
    10th problem 119
Hilbert, David ... 119, 134
homomorphism(準同型写像)
    77, 159, (257)
    example 78
    inverse (逆写像) ... 77
    λ-free (λ-自由) 77

【I】

identity
    two-sided (両側単位元) 8
index set (索引集合) 186
initial state (初期状態)
    DFA 22
    NFA 22
input alphabet
 (入力アルファベット)
    DFA 22
    NFA 22
    PDA 95
input tape
 (入力テープ) 23
instantaneous description
 (時点表示) 96
inverse homomorphism ...
 see homomorphism
inverse (逆写像) 77

【J】

justification theorem ......
 see Dedekind

【K】

Kleene closure
 (クリーニ閉包) 8
Kleene's normal form theorem (クリーニの標準形定
 理) 144
Kleene, Stephen Cole 144

Subject Index

Knuth, Donald, E. 59

【L】

λ-transition 92
Landweber, L.H. 204
language（言語）.... 2, 9
   accepted 23, 24
      by DFA 23
      by NFA 24
      by PDA 96
      by TM 150
   context-free
    （文脈自由言語）48, 168
   context-sensitive
    （文脈依存言語）... 114
   finite 17
   generated 15
   of reg. expressions .. 34
   regular（正規言語）16, 174
      pump. lem. 37
   type-0 .... 15, 82
      membership .. 178
      TM accept. ... 178
$\mathcal{L}$ closed under
   substitution 77
left recursive
  （左再帰）............. 68
Leibniz, Gottfried Wilhelm
  119
lemma
   of Bar-Hillel 72
Lemma M^+ 142
Lemma M^* 162
letter（文字）........... 7
lex 43, 44
lexical analyzer
  （字句解析器）....... 43
      generator 44
Lullus, Raimundus .. 119
$\mathcal{L}_0$
   Turing acceptable . 177
$\mathcal{L}_0$ closed under

Kleene closure 178
   product 178
   union 178

【M】

Markov, Andrey, A. 145
Matiyasevich, Yuri, V. 119
McCreight, E.M. 195, 204
meta symbol（メタ記号）33
Meyer, A.R. 195, 204
mutatis mutandis
  （必要な変更を加えて）50
μ-recursion ... 123, (277)

【N】

Naur, Peter 59
Nerode relation ....
  see relation
Nerode's theorem
  （ネローデの定理）. 31
   extension .... (237)
Nerode, Anil 31
NFA 22, (231), (235), (243)
nonterminal alphabet（非終端アルファベット）. 14
normal form（標準形）26
   Backus 59
   Chomsky 59, 70
   see also CNF
   Greibach 105, 106
   of regular grammar . 26
numbering（番号付け）146
   Gödel numbering
      definition 183
      existence 183
      isomorphic 184
   reducible 182
   universal（万能）. 147, 157, 182

【P】

pairing function

（対関数）.......... 131
palindrome（回文）9, 15, 30, 151
parser（構文解析器、パーサ）61
parse tree（構文木）.. 61
   yield（成果）........ 62
partially ordered set
  （半順序集合）........ 5
partial order（半順序）5
partial recursive ....
  see function（関数）
partition 13, 31
Pascal, Blaise 119
pattern 45
   in text 45
PCP（ポストの対応問題）157, 168, (279)
   definition 157
   solvable
   （解決可能）........ 158
   undecidable 159
   unsolvable
   （解決不能）........ 158
   1-solvable
   （1-解決可能）..... 158
PDA 60, 92, (266)
   definition 95
   deterministic . 97, (261)
   informal description 93
Post
   correspondence problem
   see PCP
Post, Emil, L. 156
power set（べき集合）2
predicate（述語）.... 129
primitive recursive 130
prefix（接頭辞）...... 34
   proper（真の）...... 34
pre-image（原像）...... 3
primitive recursion
  （原始再帰）....... 123
primitive recursive ....

see function（関数）
production
　（書き換え規則）.... 15
　（生成規則）......... 15
product（直積）........ 8
program ....
　see TM
　for S 141
　for V 141
　for Z 141
pumping lemma（反復補題）
　context-free lang. ... 72
　regular languages ... 37
pushdown automaton（プッシュダウンオートマトン）
　see PDA
Péter, Rózsa 134

【R】

random-access machine（ランダムアクセス機械）145
range（値域）........ 140
recursion theorem
　（再帰定理）....... 185
recursively enumerable ...
　see set（集合）
recursive（帰納的）.. 156
　set 156
reduced grammar
　（既約文法）....... 54
　　definition 55
　　example 56
reducibility
　（帰着可能性）..... 182
reducible（帰着可能）182
reduction（帰着）... 156
reflexive-transitive closure
　（反射的推移的閉包）. 6,
　15, 96, 111, (225)
ℛℰ𝒢 16
ℛℰ𝒢 closed under
　complement 174, (240)
　homomorphism .. (258)

intersection .... (240)
inv. homomorphism (259)
Kleene closure 17
product 17
set difference ..... (240)
transposition ..... (239)
union 17
intersection 174
regular expression
　（正規表現）......... 34
　UNIX 41
regular grammar
　（正規文法）......... 16
　normal form 26
regular language
　（正規言語）......... 16
relation（関係）......... 5
　antisymmetric
　　（反対称的）........... 5
　binary 5
　equivalence 5, 31
　　finite rank
　　（有限の階数）31, (237)
　　right invariant
　　（右不変）... 31, (237)
　　Nerode relation 31,
　　　(237)
　reflexive（反射的）... 5
　symmetric（対称的）5
　transitive（推移的）5
representative
　（代表元）............. 33
Rice's theorem
　（ライスの定理）.. 186
Rice, Henry Gordon . 186
right invariant
　（右不変）............. 31
right recursive
　（右再帰）............. 68
Robertson, E.L. 204
Rogers' theorem 184
Rogers, Hartley, Jr. . 184
rule（規則）............. 15

composition（合成）122
fictitious variables
　（架空変数の導入）122
identifying variables
　（変数の同一視）... 122
permuting variables
　（変数の置換）..... 122
primitive recursion
　（原始再帰）........ 123
μ-recursion（μ-再帰）123

【S】

Schickardt, Wilhelm　119
Schützenberger, Marcel-Paul 83
sed 41
semantics（意味）..... 34
semigroup closure
　（半群閉包）.......... 8
sentential form
　（文形式）........... 111
separated grammar
　（分離的文法）....... 69
set of strings 8
　product 8
set（集合）
　countably infinite 4
　not recursive
　　（非帰納的）........ 156
　recursively enumerable
　　（帰納可算的）..... 176
　recursive（帰納的）156
Smullyan, Raymond . 185
space measure
　（空間尺度）........ 189
stack alphabet（スタックアルファベット）...... 95
stack symbol（スタック記号）
　95
stack（スタック）...... 92
start symbol
　（開始記号）......... 14
state diagram

Subject Index

(状態遷移図) 24
state(s)（状態）....... 22
   DFA 22
   NFA 22
   PDA 95
string（文字列）........ 7
   empty 7
   length of 7
   substring 34
   transpose of 8
substitution（代入）.. 76
substring
   （部分文字列）....... 34
suffix（接尾辞）........ 34
   proper（真の）...... 34
symbol（記号）......... 7
   indivisible 7
syntax（構文）......... 34

【T】

tape-symbols
   （テープ記号）..... 138
terminal alphabet
   （終端アルファベット）14
terminal string
   （終端文字列）....... 62
theorem
   Cantor's 4, 121
   $\mathcal{CF}$ closure prop. 49
   $\mathcal{CF}$ iff PDA 111
   CNF 71
   fixed point 184
   gap-theorem 205
   GNF 106
   halting problem
      undecidable ... 154
   λ-free grammar 57
   $L(\mathcal{G}_{reg}) \notin \mathcal{REG}$. (240)
   Nerode's 31, (246)

PCP undecidable . 159
recursion 185
reduced grammar ... 55
$\mathcal{REG}$ iff DFA iff NFA 25
$\mathcal{REG} \subset \mathcal{CF}$ 49
regular expression ... 34
Rice's 186
Rogers' 184
separated grammar . 69
Smullyan's 185
$\mathcal{T} = \mathcal{P}$ 140
qrsuv 72
sru 37
$\mathcal{REG}$ closure prop. ... 17
time measure
   （時間尺度）....... 189
TM 137
   computation（計算）139
   configuration（様相）164
   definition 138
   deterministic（決定性）138
   instruction set 138
   （命令集合）....... 138
   one-tape（1 テープ）137, 138
   program
      （プログラム）..... 138
   set of states
      （状態集合）....... 138
   universal（万能）. 147, 149, 183
token（トークン）..... 43
Trakhtenbrot, Boris . 205
transition 92
   spontaneous 92
   λ-transition 92
transition function
   （遷移関数）
   DFA 22

transition relation
   （遷移関係）
   NFA 22
   PDA 95
transpose（転置、反字）8
triple（三つ組）........ 96
Turing computable
   （チューリング計算可能）140
Turing machine
   （チューリング機械）...
   see TM
Turing table 139
Turing, Alan 137

【U】

unary encoding 139
uncountable
   （非可算個）.......... 9
uncountably infinite
   （非可算無限）..... 121
undecidability
   （決定不能性）..... 154
   context-free lang.
      complement .. 172
      equality 173
      intersection ... 168
      subset 173
   general halting prob. 155
   halting problem ... 154
   PCP 159
undecidable
   （決定不能）....... 156
universal Turing machine
   see TM

【Y】

yacc 61
Young, Paul 205

和 英 索 引

【あ】

曖昧さ（ambiguity） .. 64
曖昧な（ambiguous） . 64
アクション（action） .. 61
アッカーマン-ペータ関数
　（Ackermann-Péter function） 134
アルファベット
　（alphabet） 7

【い】

1-解決可能（1-solvable） 158
1進符号化
　（unary encoded） 139
一般帰納的関数（general recursive function） . 133
一般停止問題（general halting problem） 154
意味（semantics） 34

【え】

h-正当（h-honest） . 193
NFA 22

【お】

オートマトン
　（automaton） 2

【か】

解決可能（solvable） 158
解決不能（unsolvable） 158
開始記号
　（start symbol） 14

回文（palindrome） 9, 15
書き換え規則
　（production） 15
可算無限
　（countably infinite） 4, 121
関数（function） 3
カントールの対関数
　（Cantor's pairing function） 133
カントールの定理（Cantor's Theorem） 4

【き】

記号（symbol） 7
規則（rule） 15
帰着（reduction） ... 156
帰着可能（reducible） 182
帰着可能性
　（reducibility） 182
帰納可算的（recursively enumerable） 176
帰納的
　（recursive） 156
帰納的定義
　（inductive definition） 10
帰納法による証明
　（proof by induction） 10
既約（reduced） 55
逆写像（inverse） 77
既約文法
　（reduced grammar） 55

【く】

空間尺度
　（space measure） 189
空スタックによる
　（via empty stack） 96
空文字列
　（empty string） 7
グライバッハ標準形
　（Greibach normal form） 105
クリーニの標準形定理
　（Kleene's normal form theorem） 144
クリーニ閉包
　（Kleene closure） ... 8

【け】

計算可能性
　（computability） . 113
計算量
　（complexity） 139
計算量クラス
　（complexity class） 195
計算量尺度（complexity measure） 188
計算理論（theory of computation） 1
計算論的に解決不能（algorithmically unsolvable） 155
形式言語理論（formal language theory） 1
結合的（associative） 7, 68

和英索引

決定可能（decidable） 156
決定性1テープチューリング機械（deterministic one-tape TM） 138
決定性有限オートマトン（DFA） 22
決定手続き（decision procedure） 119
決定不能（undecidable） 154, 156
決定不能性（undecidability） 154
ゲーデル数化（Gödel numbering） 183
言語（language） 2, 9
原始帰納的関数（primitive recursive function） 126
原始再帰（primitive recursion） 123
原像（pre-image） 3

【こ】

後者関数（successor function） 122
合成（composition） ... 5
恒等（equality） 6
恒等関数（identity function） 126
構文（syntax） 34
構文解析器（parser） .. 61
構文木（parse tree） .. 61
コンパイラ（compiler） 182

【さ】

最右導出（rightmost derivation） 67
再帰定理（recursion theorem） 185
最左導出（leftmost derivation） 67
最終状態（final state） 22
最終状態による（via final state） ... 96

索引集合（index set） 186

【し】

時間尺度（time measure） . 189
字句解析器（lexical analyzer） . 43
字句解析器の生成器（lexical analyzer generator） 44
時点表示（instantaneous description） 96
自発的遷移（spontaneous transition） 92
集合（set） 4
終端アルファベット（terminal alphabet） 14
終端文字列（terminal string） . 62
自由モノイド（free monoid） 7
述語（predicate） 129
受理器（acceptor） 2
受理計算（accepting computation） 23
準同型写像（homomorphism） . 77
状態（state） 22
状態集合（set of states） 22, 138
状態遷移図（state diagram） ... 24
初期状態（initial state） 22
真の接頭辞（proper prefix） 34
真の接尾辞（proper suffix） 34

【す】

推移的（transitive） 5
数論的減算（arithmetic difference） 128
スタックアルファベット（stack alphabet） .. 95
スタックきごう（stack symbol） 95
スタック（stack） 92

【せ】

成果（yield） 62
正規言語（regular language） 16
正規表現（regular expression） 34
正規文法（regular grammar） 16
生成（generation） 15
生成規則（production） 15
接頭辞（prefix） 34
接尾辞（suffix） 34
0型（type-0） 15
遷移関係（transition relation） 22
全射（surjective） 3
前者関数（predecessor function） 122
全単射（bijective） 3

【た】

対角化（diagonalization） 154, 191
ダイク言語（Dyck language） .. 82
対称的（symmetric） ... 5
代入（substitution） .. 76
代表元（representative） .. 33
高々可算無限（at most countably infinite） .. 4
単一要素の（singleton） 34
単射（injective） 3

和英索引　*219*

【ち】

値域（range） 140
チャーチの提唱
　（Church's thesis） 145
チューリング機械
　（Turing machine） 137
チューリング計算可能（Turing computable） .. 140
直積（product） 8
直接生成する
　（directly generate） 15
チョムスキー-シュツェンベルガーの定理（Chomsky-Schützenberger theorem）83
チョムスキー階層（Chomsky hierarchy） 117
チョムスキー標準形（Chomsky normal form） . 70

【つ】

対関数
　（pairing function） 131

【て】

TM
　M が関数 f を計算する（M computes function f） 139
TM M により受理される（accepted by TM M） 150
DFA 22
定義域（domain） ... 140
停止集合（halting set） 157, 177
停止問題
　（halting problem） 154
定数 0 関数（constant 0 function） 122
テープ記号
　（tape-symbols） . 138
転置（transpose） 8

【と】

等価（equivalent） 19
導出（derivation） 15
等性（equivalence） ... 19
同値関係
　（equivalence relation） 5
同値類
　（equivalence class） 5
特性関数（characteristic function） 130, 156
トークン（token） 43

【な】

長さ（length） 7

【に】

二項関係
　（binary relation） .. 5
二項和
　（binary addition） 126
入力アルファベット
　（input alphabet） . 22
入力テープ
　（input tape） 23

【ね】

ネローデ（Nerode） ... 31
ネローデ関係
　（Nerode relation） 31
ネローデの定理
　（Nerode's theorem） 31

【は】

パーサ（parser） 61
バッカス-ナウア形（Backus-Naur Form） 59
バッカス標準形（Backus normal form） 59
バー・ヒレルの補題（lemma of Bar-Hillel） 72
半群閉包
　（semigroup closure） 8

番号付け（numbering） 146
反射的（reflexive） 5
反射的推移的閉包（reflexive-transitive closure） ... 6
半順序（partial order） 5
半順序集合
　（partially ordered set） 5
反対称的
　（antisymmetric） ... 5
反転（transpose） 8
万能（universal） 147
万能チューリング機械（universal Turing machine）147
反復補題
　（pumping lemma） 37

【ひ】

非可算個
　（uncountable） 9
非可算無限
　（uncountably infinite）121
非帰納的
　（not recursive） .. 156
非決定性有限オートマトン（NFA） 22
非終端アルファベット
　（nonterminal alphabet）14
非縮小文法
　（length-increasing grammar） 115
左再帰
　（left recursive） 68
必要な変更を加えて
　（mutatis mutandis） 50
PDA ....
　see プッシュダウンオートマトン
標準形（normal form） 26

【ふ】

不可分記号
　（indivisible symbol） 7

符号関数
　(signum function) *128*
プッシュダウンオートマトン
　(pushdown automaton)
　60, 92
プッシュ (push) *92*
不動点定理 (fixed point theorem) *184*
部分帰納的関数 (partial recursive function) . *126*
部分特性関数 (partial characteristic function) *176*
部分文字列
　(substring) *34*
プログラム (program) *138*
文形式
　(sentential form) *111*
文法 (grammar) ... *2, 14*
文脈依存
　(context-sensitive) *114*
文脈依存言語 (context-sensitive language) *114*
文脈依存文法 (context-sensitive grammar) *114*
文脈自由 (context-free) *48*
文脈自由言語 (context-free language) *48*
文脈自由文法 (context-free grammar) *48*
分離的文法 (separated grammar) *69*

【へ】

閉包 (closure) *6*
閉包性
　(closure property) *17*
べき集合 (power set) . *2*

ヘッド (head) *23*

【ほ】

補集合 (complement) *53*
ポストの対応問題 (Post's correspondence problem) *157*
ポップ (pop) *92*

【み】

右再帰
　(right recursive) .. *68*
右不変
　(right invariant) .. *31*
三つ組 (triple) *96*
μ-再帰 (μ-recursion) *123*

【む】

無曖昧
　(unambiguous) ... *68*
無効化
　(cancellation) ... *191*
矛盾 (contradiction) .. *5*

【め】

命令集合
　(instruction set) *138*
メタ記号
　(meta symbol) *33*

【も】

文字 (letter) *7*
文字クラス
　(character class) .. *41*
文字列 (string) *7*

【ゆ】

有限オートマトン
　(finite automaton) *22*
有限拡大
　(finite extension) *191*
有限の階数
　(finite rank) *31*
ユークリッド原論
　(Euclid's Elements) *118*

【よ】

様相 (configuration) *164*
要素数 (cardinality) .. *2*

【ら】

ライスの定理
　(Rice's theorem) *186*
λ-自由 (λ-free) *57*
λ-自由な準同型写像 (λ-free homomorphism) ... *77*
λ-遷移 (λ-transition) *92*
ランダムアクセス機械
　(random-access machine) *145*

【り】

リード (read) *92*
両側単位元 (two-sided identity) *8*

【れ】

列挙手続き (enumeration procedure) *119*
連接 (concatenation) . *7*

【ろ】

ロジャースの定理
　(Rogers' Theorem) *184*

List of Symbols

[A]

A 138
$\rightarrow$ 14
$\mathcal{A}$ 22
$\{(ab)^{3^n} \mid n \in \mathbb{N}\}$ 75, (256)
$ap(n, m)$ 134
$m \dot{-} n$ 128
$\mathcal{A}_\cap$ .... (256)
$\{a^n b^n \mid n \in \mathbb{N}\}$ ... 29, 49
$\{a^n b^n c^n \mid n \in \mathbb{N}\}$ 53, 74
$\{a^n b^n c^n \mid n \in \mathbb{N}^+\}$ (228)
$\{a^{2^n} \mid n \in \mathbb{N}\}$ ...... (268)

[B]

B 138

[C]

$card(M)$ 2
cd .... (271)
$\mathcal{CF}$ 48
χ_A 156, 175
χ_p 130
χ_{EQ} .... (270)
χ_{LT} .... (271)
$[\varphi, \Phi]$ 188
CNF 70
cod 147, 148
$\mathcal{CS}$ 114
$\mathcal{C}_t^{[\varphi,\Phi]}$ 195

[D]

D_n 83

$\mathcal{DCF}$ .... (262)
$decod$ 147
δ 22
$\delta_\cap$ .... (256)
δ^* 23
DFA 22
$dom(f)$ 140
d_1 .... 136, (271)
d_2 .... 136, (271)

[E]

EQ .... (270)
$[x]$ 5
$\exists^\infty$ 188

[F]

f_M^n 140
$f(x_1, \ldots, x_n) \downarrow$ 139
$f(x_1, \ldots, x_n) \uparrow$ 139
$\forall^\infty$ 188

[G]

$\mathcal{G}_{reg}$ 34
$\mathcal{G}$ 14
Γ 95
$\Rightarrow$ 15
$\overset{}{\underset{\mathcal{G}}{\Rightarrow}}$ 50
$\overset{*}{\underset{\mathcal{G}}{\Rightarrow}}$ 50
$\overset{m}{\Rightarrow}$ 50
$\overset{m}{\underset{\mathcal{G}}{\Rightarrow}}$ 50
$\overset{1}{\Rightarrow}$ 70
$\overset{*}{\Rightarrow}$ 15
Göd 184
$\mathcal{G}_\mathfrak{P}$ 169

$\mathcal{G}_S$ 169
$\mathcal{G}_\cap$ .... (255)

[H]

h_L 159
$H(h)$ 193
h_R 159
H_i 55

[I]

$I(x)$ 126
INF 206, (285)

[K]

K 157
$\langle T \rangle$ 34
k_0 95
$\mathcal{K}$ 95

[L]

L 9
$L(\mathcal{A})$ 23
$L(\mathcal{G})$ 15
$L(\mathcal{K})$ 96
Λ 34
$\mathcal{L}$ 77
λ 7
$\overline{L}$ 53, (240)
LEQ .... (270)
$L(\mathcal{G}_{reg})$ 34, (240)
$L_{i,j}^k$ 35
$L(M)$ 150
L_{pal} 10, 15, 30, 151, (236)
L_{pal2} 75, 93, 96

List of Symbols

$L(\mathfrak{P}, \mathfrak{Q})$ 169
L_S 169
LT .... (270)
$\mathcal{L}_0$ 15, 82

[M]

M 138
$\mu y [\tau(x_1, \ldots, x_n, y) = 1]$ 123
μ 123
$\xrightarrow{\frac{1}{\mathcal{K}}}$ 96
$\xrightarrow{\frac{*}{\mathcal{K}}}$ 96
$M(w)$ 150

[N]

$\mathbb{N}$ 14
$N(\mathcal{K})$ 96
$\mathbb{N}$ 3
$\mathbb{N}^+$ 3
NFA 22
$\mathbb{N}^n$ 121
$N_\cap$ .... (255)
NUM 196
$\#_a(w)$ .... (248)
$NUM!$ 196, 200

[O]

$\oslash$ 34

[P]

P 14
$\mathcal{P}$ 121
$\mathfrak{P}$ 157

π_A 176
PCP 157
PDA 92
$\mathcal{P}^n$ 121
$\wp(M)$ 2
$\wp_{fin}(X)$ 95
$\mathcal{P}_\psi$ 182
$Prim$ 126
$\wp(X)$ 76
$P_\cap$ .... (255)

[Q]

$\mathfrak{Q}$ 157
$Q_\cap$ .... (256)

[R]

$\mathcal{R}$ 133
$\mathcal{REG}$ 16
$range(f)$ 140
$\leq_{red}$ 157
ρ^0 6
R_i 55
$\mathcal{R}^n$ 133
$\mathcal{R}_{0,1}$ 191

[S]

S 122, 126, 134, 140
Σ^+ 7
Σ^* 7
σ 14
sg 128
$\overline{sg}$ 128
Σ 7

$\Sigma_\cap$ .... (256)
$\sigma_\cap$ .... (255)
$S(n)$ 122

[T]

T 14
$\sim_L$ 31
$\mathcal{T}$ 140
$\Theta_\varphi \mathcal{P}'$ 186
$time(i, n)$ .... (278)
TM 137
$\mathcal{T}^n$ 140

[U]

$\mathcal{U}$ 147
$\mathcal{U}_f$ 199
$\mathcal{U}_{[\varphi, \Phi]}$ 206, (285)
$\mathcal{U}_0$ 199

[V]

V 122, 126, 140
$|s|$ 7
$V(n)$ 122

[W]

W_i .... (252)

[X]

X/ρ 5, 31

[Z]

Z 122, 126, 138, 140, 147
$Z(n)$ 122

―― 著者略歴 ――

Thomas Zeugmann（ツォイクマン トーマス）
- 1981 年 フンボルト大学大学院理学研究科
  修士課程修了（数学専攻）
- 1983 年 理学博士（フンボルト大学）
- 1991 年 ダルムシュタット工科大学助教授
- 1993 年 九州大学助教授
- 1997 年 九州大学教授
- 2000 年 リューベック大学教授
- 2004 年 北海道大学教授
  現在に至る

大久保　好章（おおくぼ　よしあき）
- 1990 年 千葉大学工学部機械工学科卒業
- 1992 年 東京工業大学大学院総合理工学研究科
  修士課程修了（システム科学専攻）
- 1995 年 東京工業大学大学院総合理工学研究科
  博士課程修了（システム科学専攻）
  博士（理学）
- 1995 年 北海道大学助手
- 2007 年 北海道大学助教
  現在に至る

湊　真一（みなと　しんいち）
- 1988 年 京都大学工学部情報工学科卒業
- 1990 年 京都大学大学院工学研究科
  修士課程修了（情報工学専攻）
- 1990 年 日本電信電話株式会社勤務
- 1995 年 博士（工学）（京都大学）
- 2004 年 北海道大学助教授
- 2007 年 北海道大学准教授
  現在に至る

英語で学ぶ計算理論
Theory of Computation　　　© Zeugmann, Minato, Okubo 2009

2009 年 4 月 30 日　初版第 1 刷発行

検印省略	著　者	Thomas	Zeugmann
		湊	真　一
		大久保	好　章
	発行者	株式会社	コロナ社
	代表者	牛来辰巳	
	印刷所	三美印刷株式会社	

112-0011　東京都文京区千石 4-46-10
発行所　株式会社　コロナ社
CORONA PUBLISHING CO., LTD.
Tokyo Japan
振替 00140-8-14844・電話(03)3941-3131(代)
ホームページ http://www.coronasha.co.jp

ISBN 978-4-339-02438-8　（新宅）　　（製本：愛千製本所）
Printed in Japan

無断複写・転載を禁ずる
落丁・乱丁本はお取替えいたします

電子情報通信レクチャーシリーズ

■(社)電子情報通信学会編　　(各巻B5判)

共通

配本順			頁	定価
A-1		電子情報通信と産業	西村吉雄著		
A-2	(第14回)	電子情報通信技術史 —おもに日本を中心としたマイルストーン—	「技術と歴史」研究会編	276	4935円
A-3		情報社会と倫理	辻井重男著		
A-4		メディアと人間	原島博 北川高嗣 共著		
A-5	(第6回)	情報リテラシーとプレゼンテーション	青木由直著	216	3570円
A-6		コンピュータと情報処理	村岡洋一著		
A-7	(第19回)	情報通信ネットワーク	水澤純一著	192	3150円
A-8		マイクロエレクトロニクス	亀山充隆著		
A-9		電子物性とデバイス	益一哉著		

基礎

B-1		電気電子基礎数学	大石進一著		
B-2		基礎電気回路	篠田庄司著		
B-3		信号とシステム	荒川薫著		
B-4		確率過程と信号処理	酒井英昭著		
B-5		論理回路	安浦寛人著		
B-6	(第9回)	オートマトン・言語と計算理論	岩間一雄著	186	3150円
B-7		コンピュータプログラミング	富樫敦著		
B-8		データ構造とアルゴリズム	今井浩著		
B-9		ネットワーク工学	仙石正和 田村裕 共著		
B-10	(第1回)	電磁気学	後藤尚久著	186	3045円
B-11	(第20回)	基礎電子物性工学 —量子力学の基本と応用—	阿部正紀著	154	2835円
B-12	(第4回)	波動解析基礎	小柴正則著	162	2730円
B-13	(第2回)	電磁気計測	岩﨑俊著	182	3045円

基盤

C-1	(第13回)	情報・符号・暗号の理論	今井秀樹著	220	3675円
C-2		ディジタル信号処理	西原明法著		
C-3		電子回路	関根慶太郎著		
C-4	(第21回)	数理計画法	山下信雄 福島雅夫 共著	192	3150円
C-5		通信システム工学	三木哲也著		
C-6	(第17回)	インターネット工学	後藤滋樹 外山勝保 共著	162	2940円
C-7	(第3回)	画像・メディア工学	吹抜敬彦著	182	3045円
C-8		音声・言語処理	広瀬啓吉著		
C-9	(第11回)	コンピュータアーキテクチャ	坂井修一著	158	2835円

配本順			頁	定価
C-10		オペレーティングシステム	徳田 英幸 著		
C-11		ソフトウェア基礎	外山 芳人 著		
C-12		データベース	田中 克己 著		
C-13		集積回路設計	浅田 邦博 著		
C-14		電子デバイス	和保 孝夫 著		
C-15	(第8回)	光・電磁波工学	鹿子嶋 憲一 著	200	3465円
C-16		電子物性工学	奥村 次徳 著		

展開

D-1		量子情報工学	山崎 浩一 著		
D-2		複雑性科学	松本 隆 編著		
D-3	(第22回)	非線形理論	香田 徹 著	208	3780円
D-4		ソフトコンピューティング	山川 尾 烈 堀 恵二 共著		
D-5	(第23回)	モバイルコミュニケーション	中川 正雄 大槻 知明 共著	176	3150円
D-6		モバイルコンピューティング	中島 達夫 著		
D-7		データ圧縮	谷本 正幸 著		
D-8	(第12回)	現代暗号の基礎数理	黒澤 馨 尾形 わかは 共著	198	3255円
D-9		ソフトウェアエージェント	西田 豊明 著		
D-10		ヒューマンインタフェース	西田 正吾 加藤 博一 共著		
D-11	(第18回)	結像光学の基礎	本田 捷夫 著	174	3150円
D-12		コンピュータグラフィックス	山本 強 著		
D-13		自然言語処理	松本 裕治 著		
D-14	(第5回)	並列分散処理	谷口 秀夫 著	148	2415円
D-15		電波システム工学	唐沢 好男 著		
D-16		電磁環境工学	徳田 正満 著		
D-17	(第16回)	VLSI工学 ―基礎・設計編―	岩田 穆 著	182	3255円
D-18	(第10回)	超高速エレクトロニクス	中村 徹 三島 友義 共著	158	2730円
D-19		量子効果エレクトロニクス	荒川 泰彦 著		
D-20		先端光エレクトロニクス	大津 元一 著		
D-21		先端マイクロエレクトロニクス	小田 光 柳中 正徹 共著		
D-22		ゲノム情報処理	高木 利久 小池 麻子 編著		
D-23	(第24回)	バイオ情報学 ―パーソナルゲノム解析から生体シミュレーションまで―	小長谷 明彦 著		近刊
D-24	(第7回)	脳工学	武田 常広 著	240	3990円
D-25		生体・福祉工学	伊福部 達 著		
D-26		医用工学	菊地 眞 編著		
D-27	(第15回)	VLSI工学 ―製造プロセス編―	角南 英夫 著	204	3465円

定価は本体価格＋税5％です。
定価は変更されることがありますのでご了承下さい。

図書目録進呈◆

コンピュータサイエンス教科書シリーズ

(各巻A5判)

■編集委員長　曽和将容
■編 集 委 員　岩田　彰・富田悦次

配本順			頁	定価
1.	(8回)	情報リテラシー	立花 康夫／曽和将容／春日秀雄 共著	234	2940円
4.	(7回)	プログラミング言語論	大山口 通夫／五味 弘 共著	238	3045円
6.	(1回)	コンピュータアーキテクチャ	曽和将容 著	232	2940円
7.	(9回)	オペレーティングシステム	大澤範高 著	240	3045円
8.	(3回)	コ　ン　パ　イ　ラ	中田育男 監修／中井央 著	206	2625円
11.	(4回)	ディジタル通信	岩波保則 著	232	2940円
13.	(10回)	ディジタルシグナルプロセッシング	岩田　彰 編著	190	2625円
15.	(2回)	離　散　数　学 ―CD-ROM付―	牛島和夫 編著／相利廣雄／朝民一 共著	224	3150円
16.	(5回)	計　算　論	小林孝次郎 著	214	2730円
18.	(11回)	数　理　論　理　学	古川康一／向井国昭 共著	234	2940円
19.	(6回)	数　理　計　画　法	加藤直樹 著	232	2940円
20.		数　値　計　算	加古孝 著		近刊

以下続刊

2.	データ構造とアルゴリズム	熊谷 毅著	3.	形式言語とオートマトン 町田 元著
5.	論　理　回　路 渋沢・曽和共著		9.	ヒューマンコンピュータインタラクション 田野俊一著
10.	インターネット 加藤聰彦著		12.	人工知能原理
14.	情報代数と符号理論 山口和彦著		17.	確率論と情報理論 川端 勉著

定価は本体価格+税5%です。
定価は変更されることがありますのでご了承下さい。

図書目録進呈◆